RYUHO OKAWA'S SPIRITUAL TRIP TO MARS

HS Press

RYUHO OKAWA'S SPIRITUAL TRIP TO MARS

Dream Interpretations
by R. A. Goal and Edgar Cayce

EL CANTARE
RYUHO OKAWA

HS PRESS

Original title: *Ryuho Okawa, Yumehandan Reading*
- R.A.Goal to Edgar Cayce no Chikyu to Uchu no Miraizu
HS Press is an imprint of IRH Press Co., Ltd.
Tokyo

ISBN: 978-4-8233-0481-1

First Edition

RYUHO OKAWA'S SPIRITUAL TRIP TO MARS

Dream Interpretations by R. A. Goal and Edgar Cayce

EL CANTARE
RYUHO OKAWA

HS Press

Original title: *Ryuho Okawa, Yumehandan Reading*
- R.A.Goal to Edgar Cayce no Chikyu to Uchu no Miraizu
HS Press is an imprint of IRH Press Co., Ltd.
Tokyo

ISBN: 978-4-8233-0481-1

First Edition

Contents

CHAPTER ONE

<Mars> Dream Interpretations

—R. A. Goal Reveals the Truth About an Unknown Martian Civilization

CHAPTER TWO

A Dream Interpretation by Edgar Cayce

—The Crisis of an Imminent Space War

*All chapters compiled in this book are originally spiritual readings recorded in Japanese and later translated in English.

EDITOR'S NOTE

From ancient times, it has been said that enlightened ones like Shakyamuni Buddha were able to freely wield six kinds of superhuman power, which are called the Six Divine Supernatural Powers: astral travel, clairvoyance, clairaudience, mind-reading, fate-reading, and *rojin* (power to surpass worldly desires at will). They are the highest level of spiritual abilities that transcend time and space and allow the user to see through the past, present, and future at will. Master Ryuho Okawa, who possesses the world's most advanced spiritual abilities, can use these powers on command to conduct various spiritual readings.

To conduct the readings compiled in this book, Master Okawa used the abilities of spiritual speech, spiritual view, time travel reading (seeing the past or future state of the target), remote viewing (seeing the situation of a certain place by sending a part of his spirit body), mind reading (reading the mind and thought of the target, even someone in a remote location), and mutual conversation (speaking the thoughts of various beings with whom we cannot communicate through conventional means).

<Backstory of the spiritual readings>

This book consists of spiritual readings conducted on the dreams that Master Ryuho Okawa of Happy Science had during sleep. It is a compilation of two spiritual readings, "<Mars> Dream Interpretations" recorded on February 17, 2021, and "A Dream Interpretation by Edgar Cayce" recorded on May 28, 2021. Both readings were conducted at the Happy Science Special Lecture Hall.

CHAPTER ONE

<Mars>
Dream Interpretations

—R. A. Goal Reveals the Truth About an Unknown Martian Civilization

Recorded on February 17, 2021
at Happy Science Special Lecture Hall.

R. A. Goal

A space being from Planet Andalucia Beta in Ursa Minor. He is one of the commanders of the space defense force and is a certified messiah. One of his roles is to serve as a religious leader, and he has the power to create an advanced civilization on a planetary scale. His current role is to protect El Cantare, who is now living as Ryuho Okawa. R. A. Goal is one of the space souls of Shakyamuni Buddha, who is a soul sibling of El Cantare, the Creator God and God of Earth.

Questioner from Happy Science

Shio Okawa*

Aide to Master & CEO

**Her professional title represents her position at the time of the recording.*

1
A Spiritual Reading on My Dream About Going to Mars

RYUHO OKAWA

Today is Wednesday, February 17, 2021. I had three different dreams between last night and this morning.

In my first dream, East Japan and West Japan had been split apart, and for some reason, I was investigating something underground with a bank employee whom I did not recognize. Since I was underground, I was probably calculating and recording how many people had committed evil in East Japan and West Japan, separately. I was doing this in something like an account book. That is when I woke up the first time, and when I fell back asleep, the dream picked up where it left off.

I must have had this dream because a big earthquake of magnitude 7.3 hit East Japan just the other day, which was a few days after the National Foundation Day (Feb 11). It struck Fukushima and Miyagi Prefectures, which are quite close to where the Great East Japan Earthquake had struck in 2011. Anyway, I must have been investigating the cause of the earthquake in my dream. I was most likely counting the number of people who had committed evil in East Japan and was about to do the same for West Japan next. I had this dream after hearing a spirit say that an earthquake of the same magnitude will strike West Japan next. So, I believe I was calculating the distribution of people who have sinned and are considered guilty from a religious perspective to investigate which areas of Japan are "permissible" for an earthquake to occur. I saw these two dreams consecutively, but they did not develop any further.

The third and last dream began around dawn, and this time, I was in space. For some reason, I had traveled to what I believe was Mars, but it did not look like how we normally imagine it. The Mars I saw had many prosperous cities similar to those on Earth. I think I was on a ship, looking at the surrounding cities from the deck. The cities resembled Manhattan or European cities; the Martian civilization was in no way inferior to that of Earth's. I had a dream where I was looking at the Martian civilization from a ship as it moved across the ocean. I was thinking, "Is this really Mars? What a surprise." I probably do not remember everything that happened in the dream, but when I started having those thoughts, I felt something pull me strongly from behind. Perhaps I was pulled by my silver cord; I was drawn back to Earth, to my physical body. Toward the end of my trip to

Mars, my heart felt a little burdened, and it continued shortly after I returned to my physical body. Perhaps I was brought back because of that. I may have traveled there as a spirit body. Anyway, it was a slightly unusual dream.

In today's evening edition of *Mainichi Newspaper*, there was an article on NASA's Mars rover, Perseverance, that was launched in July 2020. It said that the rover will land on Mars in the afternoon of February 18 in the U.S., which is the morning of February 19 in Japan. The purpose of dispatching the rover was to search for any signs of life because microbes are thought to have lived on Mars several billion years ago. The article also said that the rover will be subjected to a high heat of about 1,300 degrees Celsius (about 2,370 degrees Fahrenheit) upon entry because, although thin, Mars has an atmosphere. The targeted landing spot is located on a crater near the equator, which has a diameter of 45 kilometers (28 miles). Because there are sediments accumulated from flowing water, it is presumed that the crater was formerly a lake about 3.5 billion years ago, when Mars had a temperate climate. NASA plans to bring back a capsule containing soil from Mars to analyze and gather evidence of microbial life (in a future exploration).

We currently understand Mars as a reddish-brown, desert-like planet. It has a very thin atmosphere and is not considered a habitable place for bigger organisms. However, the Mars I saw in my dream had a civilization very similar to the one on Earth. In the space people readings I conducted in the past, I heard many times that Mars has a front-line base for space people. But in reality, we cannot see a base like that on the surface of the planet. Perhaps the surface could be fake and camouflaged, so that you can't see it, or there could be underground cities with water. I also wonder why I specifically went to Mars. I believe I was there in the morning, between 5 a.m. and 7 a.m. So, I would like to conduct a spiritual reading to interpret this dream and find out how accurate it was and what it meant.

Is Mars really a planet with a thin atmosphere, traces of water, and a little ice from the past that is only habitable for microbes, just as the current scientific research suggests? Or do humanoids live there without being detected? Or is what I saw in my dream a scene from an ancient past or the far future? There are all kinds of possibilities. So, to find the answers to all these questions, I would like to conduct a reading on the dream I had where I went on a trip to Mars.

Well then, let us begin. This morning, in my last dream, I may have traveled to Mars, but it may not have been just a dream. I strongly felt that I was really visiting Mars. I saw things that cannot be found on the surface of Mars, such as an ocean, a ship, and modern cities with many tall buildings on land. Why did I see them? As part of the dream interpretation, if there was anyone who wanted to show me these scenes, anyone with some kind of connection to my dream, or anyone who wanted to convey something to me, please come and tell me.

[*About 25 seconds of silence.*]

2
A Space Being and Its Aim of Guiding Master Okawa to Mars

Ryuho Okawa's soul was put into a cyborg and taken to Mars in a UFO

SPACE PERSON FROM ANDALUCIA β (HEREAFTER SP)

I have come from Planet Andalucia β.

SHIO OKAWA

Andalucia β?

SP

Yes. I took him to Mars this morning.

SHIO OKAWA

How are you involved with Mars?

SP

We are currently constructing a front-line base there, so I wished to take him there and have him take a look. I knew he could not stay for very long, so we were only there for about an hour.

SHIO OKAWA
Was it an out-of-body experience, or did he ride a UFO to get there?

SP
Hmm. [*About 10 seconds of silence.*] Well, I would have to reveal a secret to answer that question, so it is a bit difficult to answer.

In the case of Mr. Ryuho Okawa, we have a separate costume prepared for his soul to enter after leaving the physical body because his soul possesses a tremendous amount of energy. The costume is not exactly a Grey alien, which is well-known on Earth, but a far more refined version. It is basically a cyborg-type costume that feels like your own body once your soul enters it. So, we put his soul into this costume to make sure he could move around on Mars, and then we went there in a UFO.

SHIO OKAWA
So, he actually traveled to Mars in a UFO?

SP
Yes, he did. The Grey-type cyborg itself has no soul, but when your soul enters it, you directly experience everything

the cyborg experiences as if you are in a human body. I had him go inside a costume like that.

It only takes about five minutes to get to Mars. After we landed, I gave him a tour of Mars. The scenery he saw was certainly different from what people see from spacecraft or rockets that come from Earth.

An underground city on Mars

SHIO OKAWA

We, earthlings, can't see any cities like that when we investigate Mars or look at it through the telescope.

SP

No, you can't. You simply can't.

SHIO OKAWA

Does that mean the city is not on the surface of Mars?

SP

Well, you've heard that many space people from other planets have come to Mars, haven't you?

SHIO OKAWA
Yes. I've heard that Mars has become a front-line base.

SP
That's right.

SHIO OKAWA
I've also heard that the native inhabitants of Mars often have a more aggressive character.

SP
Yes. Anyhow, there is an underground city. There is a gigantic underground cavern with abundant water. Perhaps it would take a few more decades for earthlings to get there.

SHIO OKAWA
Then, the underground city...

SP
The city lies underground, and it is camouflaged, so that it can't be seen.

SHIO OKAWA
Do you mean there's something like a force field?

SP

Mars may only appear to be an expanse of reddish-brown land, but it actually has structures built on it. They are meant to be invisible. The city is hidden by a dome that reflects the scenery around it, so it can't be seen from the outside. You simply won't be able to see it.

Wish for Ryuho Okawa to know more about the Laws of the Universe in detail

SHIO OKAWA

What was the main purpose of taking Master to Mars this time?

SP

We believe it's time. We would like him to know more about the Laws of the Universe in detail. We have been told to guide him to various planets—of which there are countless. Unlike the overseas missionary tours on Earth, the "outer-space missionary tours" are significantly more challenging. It's also difficult to keep a record of them, so we are now practicing how this tour should take place.

He recorded a dream of traveling to another planet with former President George H. W. Bush and his wife in one of his books. He does not clearly say which planet it was, but he talks about landing on a planet with abundant grass, where cows from Earth are free to roam. It seems like he only remembered the part where the spaceship door opened, and he got off the craft and saw cows and the surrounding scenery (refer to *The Descent of Japanese Father God Ame-no-Mioya-Gami* [Tokyo: HS Press, 2022]). Anyhow, we would like him to accumulate more experiences like this to gain more accurate information about the universe. This year, we are planning to have him practice experiencing outer space whenever he has time.

SHIO OKAWA

Master said his heart felt a little burdened on the way back from Mars. Was it because he is still not used to going to outer space?

SP

He may have had some difficulty breathing because he is not used to the costume, which, by the way, is not a Grey cyborg, but a little bigger than that. The costume is made for humans, but he wasn't used to it yet. That's why we promptly returned him to Earth.

SHIO OKAWA

I see. Andalucia β is Mr. R. A. Goal's planet...

SP

Yes. That's right.

SHIO OKAWA

Are you positioned under him as his subordinate?

SP

We have to reveal what kind of beings R. A. Goal and the people of Andalucia β are, but we have not gotten there yet. For the time being, I thought Master Okawa should practice going to Mars and recording his observations first.

SHIO OKAWA

What's your name?

SP

Me?

SHIO OKAWA

Yes.

R. A. GOAL
I'm R. A. Goal.

SHIO OKAWA
Oh, you're Mr. R. A. Goal [*laughs*].

R. A. GOAL
I'm truly sorry.

SHIO OKAWA
No, I should be the one to apologize for not recognizing you.

R. A. GOAL
I'm sorry for hiding my name from you.

SHIO OKAWA
Please don't be sorry. I thought you were someone else.

R. A. GOAL
I couldn't bring myself to... In any case, I'm sorry. I was a little embarrassed to say that a commander like me took care of a job like that.

SHIO OKAWA
Please don't worry. Actually, thank you very much. I am very relieved that it was you, Mr. R. A. Goal, who took Master to Mars.

R. A. GOAL
I plan first to take him to a couple of planets within the Solar System where many space people gather. So, we will start with the dark side of the Moon, other planets in the Solar System, and then go to the exoplanets near the Solar System. Once he gets used to going there, we may take him to places that are located much further away and bring him back. This is possible if Master himself gets the hang of it. We know that you will be guarding his physical body well, so that other things can't enter his body while his soul is away.

Well, today, we were riding on a ship along the coast and looking at the city, but it seemed like he was not feeling very well, so we decided to return him to Earth. It may have been better to send him back on a spaceship, but I felt he should go back as soon as possible. So, instead of putting him in a UFO, I tugged on his silver cord and pulled his soul out of the cyborg and back to Earth. He returned from Mars to Earth in just a few seconds, so he was traveling extremely fast.

That might be the reason he felt slightly off and felt a little pain when he returned to his physical body.

In any case, an underground city does exist. Since Mars is close to Earth, it is a very convenient location to build a front-line base to monitor Earth. That's why I felt that I should show it to him first.

3
The Space People Who Have Come to Mars

What exists on Earth also exists on Mars

SHIO OKAWA
Are there any humanoid races living on Mars that Master saw?

R. A. GOAL
Yes. Well, he did not get to walk through the city. He saw the city as he traveled along the coast on a ship, as if he were gazing at Manhattan City or a European city from the coastline. He could see that there are many modern buildings on Mars and that it is no different from a modern and developed country on Earth. That was as far as his trip took him.

SHIO OKAWA
If I remember correctly, when Master returned, he said, "It seemed like what exists on Earth also exists on Mars."

R. A. GOAL
That's right.

SHIO OKAWA
He said, "It really did not seem any different from Earth."

R. A. GOAL
Yes. Well, I could not have him come to Mars just to eat sunflower seeds and send him back. We do have various things prepared for visitors. We sometimes invite guests from other planets as well, and some of them are interested in seeing Earth. In that case, we have them stay on Mars first before taking them to Earth. So, our spaceships not only carry space people who are stationed near Earth for a purpose like ours... They also carry guests who are here temporarily before going to see Earth.

SHIO OKAWA
Another thing Master mentioned was that, given how advanced Mars is, we would be no match for the Martians if they were to attack us on Earth.

R. A. GOAL
Well, Martians are indeed more advanced in UFO technology than earthlings, but Mars also houses the bases of the protector gods of Earth. Having said that, they do not have control over everything. So, to station a base on Mars to

observe Earth, you need to comply with not the international laws but with the "space laws."

Laws similar to those on Earth also exist in space

SHIO OKAWA
The other day, a space person who had "walked into" Xi Jinping also mentioned Mars...

R. A. GOAL
Didn't you have a dream about a space war?

SHIO OKAWA
Yes, but my dreams are unreliable. I think the space person controlling Xi Jinping behind the scenes said something about a base on Mars... (refer to *The Xi Jinping Thought Now* [Tokyo: HS Press, 2021]).

R. A. GOAL
Yes. It certainly does have a base there. However, the space people on Mars do not fight one another directly. When they do, it will turn into a much bigger conflict, so there is a mutual agreement not to fight.

SHIO OKAWA

I would like to ask a question based on an earthling's perspective. You mentioned that Mars has front-line bases, native Martians, and various space people from other planets making bases and living there. Am I correct to assume that Mars is similar to the world depicted in *Star Wars*, where people of various appearances live differently?

R. A. GOAL

Yes. Mars has a similar concept to nations, where different types of space people live in separate areas.

SHIO OKAWA

For example, if you, Mr. R. A. Goal, arrived from a different planet and decided that you want to build a base on Mars, can you simply claim a piece of land and start building a base there without telling anyone about it?

R. A. GOAL

No, I can't. There are space laws [*laughs*].

SHIO OKAWA

I see.

R. A. GOAL
There are all kinds of laws included in these space laws, such as "space civil laws" and "space criminal laws."

SHIO OKAWA
Really?! Such laws exist?

R. A. GOAL
We have space laws regarding war as well. There are many kinds of laws, and we base our actions on them. But I think it is still a bit difficult for you to understand them.

Current earthlings cannot easily accept that Mars has a metropolis with one million people

R. A. GOAL
The other thing is about past civilizations. Mars has had past civilizations, so I must speak about them as well at some point...

SHIO OKAWA
Did they perish?

R. A. GOAL

Yes. They were quite advanced in the past, but they perished... Eventually, I will also have to share about the civilization on the dark side of the Moon. It takes some time for earthlings to become accustomed to matters of the universe, so I can only reveal information little by little. Even if they came to Planet Andalucia β, had some sunflower seeds, and then returned to Earth, they would just wonder, "Huh? What was that all about?" and that'd be all. They would probably think they had gone to the Himalayas and practiced meditation. (Refer to *R. A. Goal's Words for the Future* [Tokyo: HS Press, 2021].)

SHIO OKAWA

OK, so the planet Mars that Master saw actually had buildings?

R. A. GOAL

Yes. He saw Martian cities that looked a lot like Manhattan or perhaps Amsterdam in the Netherlands.

SHIO OKAWA

What do you mean by Amsterdam? Were there brick buildings?

R. A. GOAL
Brick buildings... Well, I meant to say it's a beautiful city like Amsterdam.

SHIO OKAWA
Oh, I see. So, it's beautiful.

R. A. GOAL
Yes. That's what I meant.

SHIO OKAWA
Do you mean the city has European aesthetics?

R. A. GOAL
Well, there are also many modern buildings, but to reveal everything about the cities, Master must teach a

A view of Manhattan (left) and Amsterdam (right).

little more about the Laws of the Universe. Earth's level of space technology is still quite primitive. People are currently planning to send a rover to Mars, which takes a couple of days to land, and then bring back soil to analyze it for any microbial life on Mars. This level of technology is underdeveloped.

SHIO OKAWA
Earthlings think there are no life forms on Mars because there is no water.

R. A. GOAL
Actually, it has recently become known that there is some water on Mars, so earthlings now think that there might be some microbes. There are also traces of canals from the past... Martian canals are famous. There are traces of canals and lakes, so astronomers think that the water disappeared at some point by chance, but the truth is that there is still water underground. There is a whole other world expanding and being built.

Right now, the surface temperature of Mars is too high. If the atmosphere were denser and thick clouds covered the planet, temperatures could drop. But to explain this fully, I must touch upon past Martian civilizations and how they changed, so I cannot disclose it completely. In any case, we live there, and it's quite... I wonder how big it is... Well, it is a metropolis with a population of over a million space people. At the very least, it is a big enough city that would be worth visiting for El Cantare.

SHIO OKAWA
Is it kind of similar to New York, where people from various countries gather?

R. A. GOAL
Yes. Lots of space people gather...

SHIO OKAWA
So, they are mingling and living together.

R. A. GOAL
Yes. And if they do not get along, we separate them and have them live in different areas.

This is not so easy to explain to earthlings. When Master has fewer things demanding his attention, I would like to take him strategically to various places on Mars. I can inform him about where he is and such, and he can keep a record of them. I think that would be a good idea.

SHIO OKAWA

It sounds similar to "The Spiritual Diary of Emanuel Swedenborg," but this one will be more like "The Space Diary"...

R. A. GOAL

Yes, exactly. I doubt there are other humans who can keep a record like this.

SHIO OKAWA

No. I don't think it's possible.

R. A. GOAL

I doubt they can. Even if they are able to visit space, they won't be able to withstand the shock. It is also a matter of ability. We used a UFO to take him to Mars, but we didn't use one on the way back. We pulled on his silver cord like an elastic band and brought him back to Earth in just a few seconds. That means it was done at an incredibly high speed. It definitely had some impact on his physical body. So, the person traveling with us must be able to withstand this kind of shock.

Mars has an artificial sun and sky underground

SHIO OKAWA
Is there oxygen on Mars?

R. A. GOAL
Yes. We actually make oxygen ourselves since it can be made.

SHIO OKAWA
Then, you can make both oxygen and water.

R. A. GOAL
Yes, we can. We make everything: oxygen, water, and even soil to grow various crops.

SHIO OKAWA
How is the temperature there?

R. A. GOAL
The temperature is set at what is considered comfortable to help space people get used to the climate on Earth. When they come from other places, they usually rest a little on Mars before traveling to Earth.

SHIO OKAWA
Is it slightly warmer than Earth?

R. A. GOAL
Hmm... Well, we have a cooling and heating system everywhere, so I'm not sure what to say...

SHIO OKAWA
But the outside is...

R. A. GOAL
Yes. But in the underground city, we have an artificial sun.

SHIO OKAWA
Wow!

R. A. GOAL
Yes. We have a ceiling, but we also have an artificial sun and sky.

SHIO OKAWA
You can even make your own sun?

R. A. GOAL

Yes. On Earth, it may be similar to the "Venus Fort" in Odaiba city (currently called "Immersive Fort Tokyo"). We have a high ceiling, and there is an artificial sun for lighting. The sun moves according to the number of hours we have in a day and creates time. We have air and a blue sky. We built it so that it appears like we have a sky. We have water as well.

4
About the Immigration Plan of Earthlings to Mars

Leaving behind a "secret scripture" of the universe for earthlings to awaken

SHIO OKAWA
Are there any places on Mars where our messiahs—Mr. R. A. Goal, Mr. Yaidron, and Mr. Metatron—can meet and talk?

R. A. GOAL
It's not exactly a big mansion, but each one of us has our own place...

SHIO OKAWA
You have your own houses there?

R. A. GOAL
Yes. Each of us owns an embassy-like house.

SHIO OKAWA
Can you meet each other in person there?

R. A. GOAL

Yes, we can. This is technically a secret, but it will eventually be revealed anyway. I'm sure everything will be revealed. It is too bad you can't bring a photographer with you. That would be too difficult. Master Okawa will have to describe the scenes himself. There aren't many earthlings who can gain permission to enter Mars. In fact, there is no point in bringing someone who cannot withstand the shock of seeing Martian civilization. We can only invite those who have enough understanding to be able to explain what they saw in space to other people.

There is also a plan for earthlings to immigrate to Mars. I think earthlings will migrate to Mars in the next century. That is why people on Earth need to become accustomed to Mars and get used to the idea that life forms are already living there. It seems like China is now trying to make the dark side of the Moon habitable. However, they have not yet studied the many space treaties that have been set in place. There will be problems if they don't study them. The United States saw many things they shouldn't have seen on the dark side of the Moon, so they haven't gone back there since. They probably think that Mars would be safer. In any case, they need to keep researching space.

Ultimately, if humans use space people's technology entirely, they will be able to travel to and from places that are located millions of light-years away. I do not know how much time we have left, but given how much of the Laws of the Universe have been revealed so far, I am hoping to give detailed descriptions of the various stars that you currently know of within the next 10 years.

Master Okawa is being interrupted a lot, isn't he? The staff members at the Religious Affairs Headquarters (of Happy Science) are also being scolded for that now. When their minds are not peaceful and they are moving about restlessly like people working at an ordinary workplace, it becomes difficult for Master to travel and come back from space. That is why he can only travel late at night. Even then, some *ikiryo* (living spirits) come to him at midnight and interrupt his space travel. If they keep haunting him, he can't leave his body. If his body is empty, they may attempt to attack him.

SHIO OKAWA

What's more, the more he listens to the worldly concerns of people of this world, the more distant space feels.

R. A. GOAL

Right, right, exactly. Space will feel very distant. Even within the Religious Affairs Headquarters, ikiryo of staff are competing with each other for a spot near Master. So, I am not sure if this is the right time or a bad time to disclose information about space. I know I have to teach earthlings about the scientific gap between Earth and other planets and awaken them spiritually. However, if this is still too early and difficult for them to understand, I may just have to leave the teachings with Master, and he can compile it as a "secret scripture" for earthlings to hear about later.

When will earthlings start living on Mars?

SHIO OKAWA
Then, your aim is to leave behind some teachings for future generations through Master, so that they will have something to refer to when Earth advances further and they start interacting with space people in the near future.

R. A. GOAL
That's right. In the 2100s, some of the earthlings will likely be living on Mars, so there is a need to leave some reference materials behind in advance. Space people on Mars are experiencing an even more advanced civilization than the one on Earth. It would not be good if the arrival of earthlings became something bad, like an attack by Native Americans.

We are currently trying not to show ourselves, but we will eventually have to interact with earthlings at some point. So we, space people, are also traveling to Earth now to gather and analyze information.

SHIO OKAWA
If Mars has an ocean, does it mean there are fish, too?

R. A. GOAL
Ah, yes. We are, of course, conducting research about such matters as well. We have already started experimenting and managed to make various creatures.

SHIO OKAWA
Are you also hunting those creatures and eating them?

R. A. GOAL
Yes, but there are also space people who do not.

The culture shock of seeing the Martian city

R. A. GOAL

Today, we were on a fast-moving ship off the coast, and it seemed like Master Okawa was experiencing some culture shock as he was looking at the city.

SHIO OKAWA

Yes. He must have been surprised to find that Mars was so advanced.

R. A. GOAL

When he felt surprised and thought, "What?! Mars is no different from Earth?" I had him return to Earth.

SHIO OKAWA

So, it means Master's soul that is currently in this world actually traveled to Mars.

R. A. GOAL

Yes, exactly. I took him during the time when you were protecting him, and others could not interfere. This is the best time for him to come to space. He has traveled to both the Spirit World and space, so I think it is now necessary for him to keep a record of his trip.

I hope he will be able to come with me for a certain amount of time, but Happy Science as a whole is restless, and all kinds of problems are occurring way too often. He is being held back by worldly problems, so we want the whole organization to level up.

SHIO OKAWA

I see.

R. A. GOAL

These things need to be worked on.

Today, I brought Master Okawa to Mars, but it must have been very different from what he knew... I think he imagined there to be some kind of transparent dome-like structure on the surface of Mars that allowed people to live inside and kept them invisible from the outside. I think he only expected to see something like a colony. So, when he saw that there was a civilization on Mars that was no different from Earth's, he must have been a little shocked. In a sense, Martian civilization has more advanced scientific technology than that on Earth.

Tens of thousands of "graduates of Earth" are living on Mars

SHIO OKAWA
Which race or species of space people are currently the majority on Mars?

R. A. GOAL
Well, there are various types. I don't know the exact distribution, but many are former earthlings or "graduates of Earth." Some former earthlings come to Mars as a preliminary step to being reborn on different planets after completing their reincarnation cycle on Earth. Some of them have actually become Martians. Former earthlings will eventually go on to other planets in the universe, but before they can do that, they stay on Mars to learn about what kind of planets there are in the universe. After that, they choose a place that is suitable for their next soul training. For this reason, we get some new graduates from Earth every year, and there is even an "Earth village" on Mars. In the near future, if Master wants to see a particular area on Mars, I can arrange that, too, as long as the inhabitants there are willing to cooperate.

SHIO OKAWA
Mr. R. A. Goal, do you wear a costume when you are on Mars?

R. A. GOAL
Well, yes [*laughs*].

SHIO OKAWA
So, you do. Is it because the environment on Mars is different from the one on your home planet?

R. A. GOAL
Well, the costumes we wear to board the spaceship are different from the ones we wear to live on Mars.

SHIO OKAWA
What kind of costume do you wear when you live on Mars?

R. A. GOAL
Hmm. It's made based on everybody's average, so that it is comfortable. We also have different clothes for outdoors and indoors. The indoor clothing is made to fit us most comfortably, but the outdoor clothing is a little different because it is what we wear to interact with other space

people. However, the outdoor costume does not have to follow as strict a criterion as the spaceship uniforms.

I would like to tell you in more detail but... Well, I will eventually invite you two together if your sleeping cycles are in sync. However, if you are visiting another place (in your sleep), you will not be able to travel together. This is how it is. Today, I briefly showed Master Okawa a prosperous coastal city on Mars from the ocean.

SHIO OKAWA
If there is an ocean, are there rivers too?

R. A. GOAL
There are some inland. The city is man-made, you see.

SHIO OKAWA
Has it been designed based on the cities on Earth?

R. A. GOAL
Yes, yes, exactly. There are tens of thousands of graduates of Earth who are living here. But they do not stay here for too long.

SHIO OKAWA
They will soon be traveling to a different planet.

R. A. GOAL
Yes. On Mars, they do some research on the place they will go next.

SHIO OKAWA
Does that mean Mars mostly functions as a point of transit?

R. A. GOAL
Yes. It is located in the most convenient place for a front-line base. There are other places, too, but... Hmm. A significant portion of the dark side of the Moon has been taken over by evil beings, so we have to be a little careful about where we set up camp there. But as for Mars, we have greater rule over it.

It is time to further deepen the Laws of the Universe

SHIO OKAWA
Then, we will consider this trip to Mars as an introduction. If there are other opportunities for Master to experience this kind of thing...

R. A. GOAL
Yes, you're right. As long as you stay interested in these topics, you will be able to record his experience.

SHIO OKAWA
I understand. Thank you.

R. A. GOAL
Also, if you want me to take you with him, I can do that. But I hesitate because it could create an unguarded moment on your side (in the earthly world).

SHIO OKAWA
You're right. I guess I'd better stay.

R. A. GOAL
Right. If you go with him, there will be no one to protect him. During late at night until around dawn, there are fewer people in Taigokan (the holy temple where Master lives), so there is less interference. But that is also when spirits from the outside sometimes intrude. If you come along with Master, no one would be there to expel those beings and their thought waves. Something needs to be done about these issues. In particular, the ikiryo of his former family members, current family members,

and former secretaries often come at night and demand things from Master for their own self-realization. In these cases, his soul cannot leave his body. Some evil spirits occasionally come to threaten him as well.

SHIO OKAWA
What you basically mean is that all of the staff must step up and strive to attain higher enlightenment.

R. A. GOAL
That's right. People's current level of enlightenment is insufficient. You need a little more...

SHIO OKAWA
You mean that, in other words, we disciples are still at the same level as ordinary, non-religious people.

R. A. GOAL
If the disciples think it is enough to guard Master physically from outsiders, they are greatly mistaken. I would like them to have a higher awareness regarding this matter. They must realize and have the mindset that protecting Master is truly a 24-hour operation. Because they aren't aware of this, they don't hold enough power now.

In any case, today, we went to Mars. He may remember other things that left an impression on him, since we went to various places. We went to the Spirit World as well, of course. As long as you remember what Master said about his dream, you can ask him to conduct a reading like today and find out who was responsible for such experiences.

SHIO OKAWA
I understand.

R. A. GOAL
Yes. Please think of today as a greeting from me.

SHIO OKAWA
OK. You gave us a very valuable talk.

R. A. GOAL
Hopefully, Master will forgive me for taking him once he finds out that it was me who took him.

SHIO OKAWA
We feel very safe with you.

R. A. GOAL

Right. We won't do anything evil, so you don't have to worry. But if he were taken away by a Reptilian, you never know what would happen. If he were not to return, his physical body on Earth may stop breathing. This can happen.

SHIO OKAWA

We must be careful not to let that happen.

R. A. GOAL

It would be worrisome, wouldn't it? If someone rushes to take him to the hospital after finding him not breathing, it would be even more dangerous. That is why we cannot give him a long tour. We can only take him during his REM cycles.

SHIO OKAWA

We need to do better and protect him securely...

R. A. GOAL

Yes. Please think of a way to do so. Please tell the secretaries to think that they also have work throughout the night. What's more, please tell them not to bother Master and you about trivial matters.

SHIO OKAWA
Yes, you're right.

R. A. GOAL
The time has come to further deepen the Laws of the Universe.

SHIO OKAWA
The entire Sangha must aim not to trouble Master so that he can do that.

R. A. GOAL
Well, the staff in the General Headquarters and *shibu* (Happy Science branches) still have a long way to go, so I think it will be very tough. Perhaps, this reading may also become a "secret law." But what must be left must be left. Otherwise, he won't be able to fully teach the Laws of the Universe.

SHIO OKAWA
Yes. Thank you very much.

R. A. GOAL
That's all for today.

SHIO OKAWA
Thank you.

CHAPTER TWO

A Dream Interpretation by Edgar Cayce

—The Crisis of an Imminent Space War

Recorded on May 28, 2021
at Happy Science Special Lecture Hall.

Edgar Cayce (1877-1945)

An American prophet and a faith healer, who is widely known as the "Sleeping Prophet" and the "Miracle Man" of the 20th century. He conducted over 14,000 psychic reading sessions while asleep, recommended treatments for illnesses, gave advice to solve people's life problems, and revealed the truth about ancient civilizations. Happy Science reveals that the core consciousness of his soul is Archangel Sariel, who is the head of the healing spirit group (Refer to *The Laws of Eternity* [New York: IRH Press, 2024]).

The questioner is referred to as "A."

1
A Strange Dream About Being in New York and Then in Outer Space

Walking through New York City as an expatriate employee for the second time

RYUHO OKAWA
Let us begin the "Dream Interpretation."

This morning, on May 28, 2021, I had a dream between 4 a.m. and 5 a.m., so I would like to interpret it. This time, we will be asking Mr. Edgar Cayce for his assistance.

In my dream, I was about 30 years old and living in New York City as an expatriate employee for the second time. I seemed to be strolling around the city, looking for a new place to move into. The trading company I worked for seemed to own some kind of an apartment building, so they suggested that I use one of their rooms. I walked through New York City and was looking at the apartment building, but its external walls were covered in black graffiti, much like what you used to see in the New York City subways back in the day. Seeing this, I remember saying, "Oh, this is no good. The market price of this building will depreciate if it's left as is, so we won't be able to sell it or rent it for a good price. This kind of problem happens because no one is taking responsibility for this building." I think I was with other people as well, so I told them, "Put up a note saying, 'The property of the General Affairs Department.'" I remember saying these words in my dream.

A dream about investigating a space-slug-like object on a planetary satellite

RYUHO OKAWA

At first, I was in New York City, but then my wife, who was sleeping in the bed next to mine, coughed a few times. That was when the location in my dream suddenly changed from New York to outer space. I thought, "Huh? I was just in New York, but now I'm in space." I was no longer on Earth but somewhere else. I wondered if it was a different planet, but I knew a planet would have some kind of atmosphere. This place had no atmosphere, so I think I was on a planetary satellite or a large asteroid, especially because I remember standing on a rocky surface. However, I could not tell how big this celestial body was. I was wearing a spacesuit. A few other people were also there, so I wasn't by myself. It seemed like we were investigating something.

In front of me, there was a big, oval, rod-shaped object that was a few dozen meters long, so I wondered what it was. To me, it looked like a "space sea cucumber" or a "space slug," but either way, it was not moving. It did not seem like just a rock based on its appearance. It seemed to be a few meters in diameter and about 20 meters (about 66 feet) long. I was still not sure if I was on a planet. The height of this slug-like object was shorter than mine. On the other side of this "space sea cucumber" were a few people in spacesuits helping me with something. I myself had a beam gun or a laser gun and was trying to slice through this space object crosswise. Perhaps one of the people on the other side was also holding something, and we were all trying to slice it like we would using an electric saw.

In any case, we first cut this rock-like object crosswise using a fine orange laser. It felt like I was cutting a dried persimmon with a knife. After slicing it, the cross-section looked like a rock embedded with garnet. There was blue and white light... blue with a white glow shining out of it. I'm not sure if the entire object was filled with garnet, but it looked like they were emitting blue light. So, it obviously was not an ordinary object.

Conducting a reading to identify the object I saw in outer space

RYUHO OKAWA
I wondered what this sea-cucumber-like thing might be, so I cut another slice crosswise. That was when my wife coughed a few times again, which brought me back from outer space to reality and into my bed. I woke up before I could identify what the object was. I immediately talked to my wife about what I had just seen in my dream and spoke to her about it again after I got out of bed. I think I remembered the dream accurately. However, it was, indeed, a strange dream, and I regret not figuring out what the object was. So, I would like to conduct a reading to interpret the dream. I got the sense that the city of New York was some kind of camouflage, and that in reality, I was in outer space from the very beginning. I would like to conduct a reading and have Mr. Edgar Cayce assist me in interpreting it. If you have any questions, please feel free to ask.

Now, let us conduct a dream interpretation reading and figure out exactly what I saw in my dream this morning.

[*About 15 seconds of silence.*]

2
What I Saw in Space During My Out-of-Body Experience and Time Travel

What the two dreams mean and the connection between them

EDGAR CAYCE

In the beginning of Master Okawa's dream, he was in New York City, so there is no question that his soul had traveled somewhere during his sleep. I believe he was having an out-of-body experience. He was walking on the streets of New York to look at an apartment owned by the company he worked for, but it was covered in black graffiti. Then, he gave orders by saying, "This is going to depreciate the value of the building. We will have to lower the price when selling or renting this building, so clean up the graffiti. Also, you must assign someone to be responsible for this." This is showing what he could have experienced in a few years if he had continued working as a businessman at a trading house. But in reality, this is also merely a convenient way of showing that he had found and acquired some kind of "residence." So, his "residence" was expressed as an apartment building, but its exterior was covered in black stuff that would diminish the value of the apartment and make it worthless. This part needs to be analyzed further to reveal what the apartment building was truly representing.

Then, his wife coughed a few times, and all of a sudden, the dream was about outer space. This shows that Master Okawa was having a dream within a dream. Even though it seemed like he was in New York, he wasn't. In truth, he was in space and on a planet or something smaller than a planet. He clearly described himself wearing a spacesuit, so I suspect that there was no oxygen on this small celestial body and there was also something strange. He then tried to cut the strange object to see what was inside. So, I think that the apartment building in New York and the strange object that was sliced open were one and the same but merely appeared differently on the outside. I presume that these two things were in fact the same thing. This means we can assume that this object was man-made, and he was investigating it to see if anyone was living inside it or on board. It was a flying object from space that landed on a celestial body and was immobile. So, I think that he and some other people were investigating what this object was. They were examining whether it was an enemy or an ally, or whether it had become stranded.

What was inside the mysterious object?

EDGAR CAYCE

Master Okawa said that he was cutting the mysterious object using a laser beam. This likely actually happened in space as part of his investigation. It must have been hard to tell what the object was unless it was sliced open... Since it was not moving, it could have been a living organism that had already died or perhaps some kind of vehicle. I think he wanted to identify what it was exactly. As he sliced through it with a laser, a blue light beamed out along with some white light, revealing that it wasn't solid to the core. It was not completely full inside. That was when he woke up and returned to Earth.

Let us analyze this mysterious object more closely. In reality, there was no one inside the object, so although he initially thought it could be a spaceship, it was some kind of space organism instead. There is no doubt about that. I am not sure if it was a big space slug or a space sea cucumber, but in any case, it was about four to five meters wide, about 20 meters long, and round at the ends. I believe it was a carcass of something like a space slug. Seeing how the blue and white light glowed from the inside after it was sliced open, it may have been a very special creature that fed on

An image of the "space slug" in outer space.

something like ore. The shining substances that were found inside are most similar to crystals. The creature likely fed on something that resembled pale blue crystals. What this means is that there are organisms in space that live off crystal-like things. I suspect that this creature died from some cause, and its carcass drifted and ended up on a planet-like body, so Master was investigating this mysterious creature. I do not know whether we should consider it as a space person, but nonetheless, it is a big space organism. It was a big slug-like lump that did not even have eyes or a nose. Creatures like this do exist in space.

When, who, and why were people exploring space?

EDGAR CAYCE
It seems that a few people were investigating this creature because it was rare. I'm guessing that this dream was a scene from the future. Master said that he was wearing a spacesuit with a glass face shield and saw other astronauts in white spacesuits as well. So, I think he was seeing a scene from the latter half of the 21st century. He likely saw earthlings investigating a space organism that they found during a space mission, so the other people were probably also earthlings.

The investigation consisting of earthlings was trying to find out what the creature was.

QUESTIONER A (HEREAFTER A)

Does that mean Master's soul actually traveled to space?

EDGAR CAYCE

That is a little difficult for me to tell... However, it does seem like he was actually there, so perhaps he was looking at a scene of a space exploration project in the future that has something to do with Happy Science. Japan's current level of interest is in *ryugu-no-tsukai* (the Japanese name for oarfish)... I mean the asteroid Ryugu. Japan has sent a rover to Ryugu to bring back a rock sample. The U.S. sent a small rover to Mars to investigate Mars' composition. What Master saw in his dream was a more advanced version of these space expeditions. Astronauts had actually traveled to space to investigate various things. I suspect the mysterious object they were examining was a fossil of an organism.

A

Where is this planetary satellite, which Master found himself on, located? Is it close to the Moon or Mars?

EDGAR CAYCE

Well, let me see... Hmm... [*About 10 seconds of silence.*] It seems to be close to Jupiter. It is either a satellite of Jupiter or... it might be a big debris that had broken off from one of the satellites of Jupiter. Those astronauts discovered something and went to investigate it. When the creature that Master saw was alive... Hmm...

By the time people discovered this creature, its insides were already hardened, but when it was alive, it probably carried an extremely strong liquid acid that resembled gastric juice. This liquid allowed the creature to digest the ores it ate. In any case, it is one of the organisms that live in space. It might not be called a space person, but it is no doubt a space organism. It seems that people had discovered it before it became a fossil. This is...

A

Has this organism come to Earth yet?

EDGAR CAYCE

I believe that will happen in the 2070s, in about 50 years. So, it seems like Master went there. He thought that he had gone to New York, when he had really gone to space. This means he can travel to outer space as if he were taking an airplane to go to New York.

A

So, Master had traveled to outer space. I see.

EDGAR CAYCE

It means he could go to space to carry out an investigation, as if he were taking a space shuttle. And cutting... I think that the astronauts actually carried laser guns, and Master saw them severing the organism. I believe they are at the stage of investigating various things, such as signs of life on various planets within the Solar System and their satellites. This might have been a future vision related to Happy Science.

A

I see. Then, was Master foreseeing them?

EDGAR CAYCE

Yes. I think he time traveled and watched the scene through the eyes of someone else. Perhaps, his descendants have something to do with it.

A

Oh. You could be right.

EDGAR CAYCE

Yes. The Faculty of Future Industry (at Happy Science

University [HSU]) seems to be making more progress by then.

Is the mysterious organism alive in the present?

A

Does that organism exist on Earth in the present (2020)? I wonder what it could be. Since you said that it was soft and fluid on the inside when it was alive but solidified into something like crystals after it died, I thought it might also exist on Earth and be discovered in the form of crystals after it fossilized. That's what I thought for a second, but maybe not?

EDGAR CAYCE

It could've been moving when it was alive. It was probably an organism that could fluidly change its shape.

A

But it cannot be classified as a space person...

EDGAR CAYCE

It's a life form or an organism.

A

I see.

EDGAR CAYCE

It doesn't have the cognitive awareness of a space person, so I would call it an organism.

The question is, what is the purpose of this organism existing? Some organisms could be feeding on ore because food is scarce in outer space. In any case, this is what the astronauts were trying to investigate. They were also looking into the past influence space had on Earth. I think they were researching various things, such as the origin of life on Earth in the things that came from space. So, while they were investigating all those things in space, they came across a mysterious, subfossilized organism. If it had landed on Earth alive, perhaps it would have been considered a new space monster [*laughs*].

A

[*Laughs*] I guess so, seeing as it's rather big.

EDGAR CAYCE

Right. If it were moving and changing shape as it crawled on the ground like a slug, I'm not sure what it would be eating on Earth, though. I think it carried very strong acidic gastric juice.

A

Oh, that's how people depict aliens sometimes. They excrete some kind of liquid that melts people or things on Earth.

EDGAR CAYCE

Yes, that's right. I don't know when the organism stopped eating. But given that the creature was partly fossilized in Master's dream, I would say that it is likely alive in the present, in the year 2021.

A

Perhaps it is orbiting Earth.

EDGAR CAYCE

The creature may still be alive now.

A

Does it have a soul?

EDGAR CAYCE

Hmm... When the astronauts were investigating it and slicing it with a laser, there was no sign of life, so the soul had probably already left its body. It might have had a will and may have been carrying out biological processes before that. At one point, they were investigating it, thinking that it was some kind of a vehicle, but they had concluded that it was a partly fossilized creature, which is basically a carcass.

3
An Undiscovered Energy Source and a Space War

How the partly fossilized organism looked 50 years before

A

Is it possible for you to speak with the consciousness of the fossilized organism?

EDGAR CAYCE

Hmm. As I said earlier, it should be in existence now.

A

Oh, that's right.

EDGAR CAYCE

It's alive now and should currently exist on one of the satellites of Jupiter or somewhere else. Hmm... Let's see if I can contact it... Hmm. [*About five seconds of silence.*] I wonder what it will look like if I rewind time by 50 years. Hmm. It looks similar to a snail without its shell.

A

That sounds like a slug.

EDGAR CAYCE

You could call it a slug, too. It has two tentacles. The one that was discovered on the satellite no longer had them, but given that it has tentacles, it must have eyes and sensory functions.

It doesn't seem to have a facial feature, but the front seems to open, so it appears to have something like a mouth.

A

What color is its body?

EDGAR CAYCE

The color was... Hmm... In its live form, it has a reddish-brown body, like an adzuki bean, and has a striped pattern. It has horizontal stripes.

A

Then, the surface of its body must not have been originally hard like rock...

EDGAR CAYCE

I think it was hard enough to be able to defend itself, but it does not seem to be slimy. It is an organism but...

A

It sounds more like a sea cucumber that lives on land.

EDGAR CAYCE

Right. But I can't tell what it feeds on.

A

Does it live by itself?

EDGAR CAYCE

No, it should have companions. This is some kind of... [*About five seconds of silence.*] Let's see if there are any similar organisms on Earth... Hmm... Based on its shape and size, there is nothing like it on Earth.

A

What is the organism doing now as you are watching it?

EDGAR CAYCE

It's crawling.

A

So, it's crawling on something near Jupiter?

EDGAR CAYCE
It's crawling on a satellite of Jupiter.

A
It's crawling on the surface of a satellite.

EDGAR CAYCE
Yes. Hmm... It could also be some kind of cyborg that collects ore and converts it into something else inside. It is unclear as to what purpose it serves, but it is at least 20 meters long. It can swallow rocks through its mouth, digest them in its stomach to extract particular components, and discard the rest as waste. That's the kind of function it has.

The work of the space slug

A
Can you possibly talk to the consciousness of that organism? Or is it too hard to do that?

EDGAR CAYCE
I don't know if it can talk by the looks of it.

A

Can we try?

EDGAR CAYCE

OK, then, let's give it a try.

A

Great.

EDGAR CAYCE

Space Slug, Giant Space Slug on a satellite of Jupiter, can you talk to us? Giant Space Slug, can you talk to us? [*About 20 seconds of silence.*] It seems like it is extracting an energy source that has yet to be discovered by humans.

A

Is the ore merely a source of food for the organism, or is someone deliberately making the organism find the ore?

EDGAR CAYCE

In this particular time in the future, people no longer use petroleum and coal as an energy source, as you do on Earth now. People no longer use resources like that. Earth will enter this stage in the near future.

In outer space, there are sources of energy that humanity has yet to discover. One of them is found in space rocks. Energy can be extracted from a particular type of ore. Hmm... Let's see... The space slug I'm looking at feeds on ore, extracts certain components, hardens them, and excretes them—this is the kind of work the slug is doing. It is an organism, but at the same time, it functions like a cyborg, so someone could be using it.

A

Then, that organism is technically doing work.

EDGAR CAYCE

Yes, it is. There are no other food sources in outer space, so it eats ore and extracts a substance that can be turned into a source of energy... The extracted components look like crystals, quartz, or something with a blue hue. I don't know what exactly these blue crystals are, but I suspect that they can be converted into energy somehow. This is an undiscovered source of energy on Earth. If energy can be extracted from these crystals, that would make space travel much easier.

A

Do you think it can be used in UFOs? I mean, as a source of fuel.

EDGAR CAYCE
Hmm... The main source of energy that UFOs use is still unknown to humans for the most part. I think rocks are potentially a widespread source of energy in space. If people can extract mineral resources from rocks and convert them into energy, they will be able to replenish their energy supply at various places.

It is impossible to space travel continuously using fuels like petroleum and coal, so there is a need for something else... That is why space slugs are sent to land somewhere and ingest energy sources whenever it is necessary. They are most likely sent from a spaceship and supply the energy back to that ship. In a way, they are digging for "space coal."

A
Wow. That is impressive.

EDGAR CAYCE
Hmm.

Who is dispatching the space slugs?

A
Here's the last question. Can you contact the spaceship?

EDGAR CAYCE

Let me ask. Who is dispatching the space slugs? The one dispatching the space slugs... There must be someone or something out there. They should currently be flying in space. Who is dispatching the space slugs? Who is extracting the undiscovered source of energy? [*About 15 seconds of silence.*] I can see two gigantic cogwheel-like objects turning.

A

In space?

EDGAR CAYCE

Hmm. I can also see something that resembles a cannon.

A

A cannon...

EDGAR CAYCE

Hmm... I can see something like the Space Battleship Yamato. It is another type of battleship. I think it is a vehicle that is equipped to fight battles in space.

A

So, are you seeing a UFO or a spaceship now?

EDGAR CAYCE

Yes. Umm... It's a combat spaceship. Those that you usually see are often round or disc-shaped, but this is a spaceship made for battle. It isn't disc-shaped.

A

I see.

EDGAR CAYCE

This type is a little elongated and carries offensive weapons. Perhaps the energy source is for these weapons. I wonder if that is why space slugs are collecting energy sources.

The identity of the mysterious spaceship

A

Then, I would like to take the liberty of asking what kind of space people this spaceship belongs to.

EDGAR CAYCE

Which space people does this space battleship belong to? I can also see something like a cannon. Who does this spaceship belong to? [*About 15 seconds of silence.*] Oh, so this is a spaceship designed to invade Earth.

A

So, does that mean the cannon is aimed at Earth?

EDGAR CAYCE

No, it isn't. The cannon is there to shoot down and destroy things that are launched from Earth.

A

From Earth...

EDGAR CAYCE

Yes. So, it means this spaceship was built to destroy space shuttles and artificial satellites launched from Earth.

A

What! Are they planning to destroy everything that comes from Earth, regardless of whether it is launched from the U.S. or China?

EDGAR CAYCE

Hmm... Well, they are here to spy on Earth now with the intention of invading it. What's more, it also seems like their main force will be arriving later. The spaceship belongs to an advance team that was sent for reconnaissance, so it carries offensive weapons, and flying saucers, of course, in case they need to use them. But it is stationed in outer space instead of on Earth. It is in charge of intercepting things launched from Earth. This must be part of a space war. I think the space people in this ship are preparing for the next space war. This spaceship is already out there in space now.

A

Well, given that Mr. Yaidron and Mr. R. A. Goal have been visiting us on Earth so often, we expected that there would be enemies.

EDGAR CAYCE

I think the ones in the spaceship are enemies. But it seems like they have come from a very far place.

A

Should we not talk to them, seeing as it could be dangerous?

EDGAR CAYCE

Hmm. I think they are working under Xi Jinping X (the evil space person influencing Mr. Xi Jinping).

A

Why am I not surprised?

EDGAR CAYCE

I get the sense that they have thoroughly thought about what they can take from Earth. They are probably thinking about what they can steal from Earth and make their own. That is why they have a space battleship. It certainly looks like a ship.

A

What they are thinking of doing reminds me of mergers and acquisitions (M&A) of companies.

EDGAR CAYCE

Oh, that's true.

A

Do some space people wander around space, carrying out an "M&A of planets" by exploiting all things that are useful to them until the planets are fully depleted?

EDGAR CAYCE

Well, those space people are trying to become the "new gods." I think they are trying to make a "divine descent" onto Earth to become gods and rule over the earthlings.

A

I see.

EDGAR CAYCE

Yes. They seem to be a different kind of space people from the Reptilians.

A

Does that mean Xi Jinping X is not your average Reptilian?

EDGAR CAYCE

No. He is a bit of a different type.

A

Is his type of Reptilian more intelligent than the typical Reptilian?

EDGAR CAYCE

They are different from the Reptilians that came to Earth a long time ago... They are likely the type that uses cyborgs extensively.

A

I see. Do they use AI to try to turn humans into machines?

EDGAR CAYCE
Even if they were to invade Earth, they would rarely come down to Earth themselves. Instead, they would send cyborgs. That is how they basically take over a planet. Mr. Yaidron is currently defending Earth against them, but I don't think he has told you about them yet.

A
You're right. He hasn't mentioned them in such detail...

EDGAR CAYCE
He hasn't said anything because it would worry you all. In any case, I think Xi Jinping X is the one who is sending enemies who sometimes come to Earth. I believe a space war is about to begin.

4
A Battle Over Territory in Space Has Begun

A space being pulling the strings behind China's ambition to dominate Earth

A

China...

EDGAR CAYCE

Yes. They have gone to Mars.

A

... has gone to Mars.

EDGAR CAYCE

In space, a battle over territory...

A

It has started, hasn't it?

EDGAR CAYCE

A battle over territory has already begun. China has gone both to the Moon and to Mars. So, countries on Earth are looking to claim territory in space; the battle has already begun.

A

I bet Japan has no idea about this.

EDGAR CAYCE

Japan is completely behind. They have totally fallen behind because it has been too peaceful. The best they can do is extract some components from a small asteroid. In contrast, China is already looking to fight battles over territory in space. If there is a place in outer space that has resources, they are thinking of making it their own territory as soon as they can.

Colonizing Mars is... Countries have already started racing each other to colonize Mars and the Moon sometime during this century. But there is definitely something that is lurking in the background that will creep in after they are colonized.

A

It's just like the evil Galactic Empire in *Star Wars*...

EDGAR CAYCE

Right. I think these space people are planning to toss humanity into space and take over Earth for themselves. In other words, the "losers" will be driven out. The basic stance they take when conquering others is to let those who obey or worship them as gods survive, while they expel the rest. This is why China wants to dominate Earth.

A

So, what China is saying is basically the same as what these space people in the background are thinking.

EDGAR CAYCE

I think China intends to drive everyone who considers them an enemy out of Earth. So, those who lose against them will have no choice but to escape to outer space.

A

Then, that means the ones who fight against them would be the "Resistance." I see.

EDGAR CAYCE

Right. It is a battle over who wins. And there are definitely space beings pulling the strings behind the scenes in all of this.

A

Are they pulling the strings because they no longer have a planet of their own?

EDGAR CAYCE

Yes, they do, but it's very far away.

A

Then, does that mean they have a desire to conquer the entire universe?

EDGAR CAYCE

Well, it is natural for anyone to aspire to embark on a great voyage to explore a place where there is life. So in reality, different space people are all dispatching many investigative units. The battle over territory in space is happening among space people, too. Many of them want to expand their territory in places where the inhabitants agree with their beliefs and will obey them. Space people are fighting against each other as well.

What this all means is that there is a possibility of war happening on Earth, much like the one that happened on Planet Beta (Planet Zeta). On Planet Beta, Mr. Yaidron and his allies won. But some of the enemies escaped and came to Earth, which is why Mr. Yaidron is chasing them down. Well... It is a very difficult situation.

The skirmishes of a space war have started

EDGAR CAYCE

In that sense, the battle may not just take place on Earth; the next battle will be a space war. Even when fighting a war on Earth, it is important to shoot down the enemy satellite as soon as possible. But next, people will need to be prepared to shoot down any space shuttles that may be used to attack the satellites in space. I suspect that the U.S., China, and Russia are already thinking this in advance.

It is actually like the world of *Star Wars*. We have already entered such a stage. There are space people who possess a base within the Solar System, and some of them are on your side, while others are not. So, what you are currently looking into is one of the skirmishes that will lead to a space war; it is a neck and neck race on Earth between the so-called "space alliance" or "planetary alliance" and the "invaders."

A

A neck and neck race...

EDGAR CAYCE

Yes. Space people haven't started their battle in a way that you can visibly see yet, have they?

A

No.

EDGAR CAYCE

Space people on both sides are making sure that they remain inconspicuous. Having said that, there have been all kinds of strange incidents, wars, or accidents happening around Earth, haven't there? For example, there were mountain fires in Australia and in the Midwestern United States that were quite unnatural. I think it is best to question whether those fires really started naturally. Like virus warfare, these incidents may have been deliberately started to weaken those countries.

A

I see. According to Ame-no-Mioya-Gami (the Japanese Father God), who gave us a talk the other day, the virus war has already started, and we are actually in the midst of it.

EDGAR CAYCE

Yes, that is certainly what's happening. Currently, viruses originating from Earth are mainly being spread, but if they become ineffective as the vaccines get distributed, viruses originating from space may be introduced next. This will lead to a very tough situation.

A
I see. Mr. R. A. Goal told us recently that evil space people might soon introduce another mutated virus strain.

EDGAR CAYCE
Indeed, they might. The virus will keep mutating. Japan is saying that it will combat the virus with vaccines from the U.S. or the U.K., but even with the vaccines, the virus will keep mutating, and the vaccines will become useless. Although the vaccines may seem to work at first, they will stop working the next time around. These incidents are the skirmishes of a proxy war that the evil space people are fighting.

Energy source of Earth in 2050

EDGAR CAYCE
The other thing is that these evil space people can make all kinds of incidents look as if they are natural phenomena... This may be the case for the recent mountain fires as

well. There really are weapons that can cause earthquakes. Earthquakes, tsunamis, and hurricanes are certainly natural phenomena, but they can also be induced through artificial means.

Right now, there are people who are after Earth's natural resources, and they are trying to control Earth by manipulating earthlings with the idea of lowering Earth's temperature and preventing global warming. To me, it seems like they are trying to bring down the level of their enemy's civilization. If earthlings are not able to use petroleum, coal, natural gas, or nuclear power, they would have to return to a primitive way of life. Solar power also becomes an unreliable source of energy when the skies are cloudy.

A

This seems ironic. Young people like Ms. Greta are protesting ardently to reduce CO_2 emissions, but they are the ones who are most dependent on electricity.

EDGAR CAYCE

Yes, that's true.

A

It looks like they want to get rid of the things they themselves rely on.

EDGAR CAYCE
You can generate wind power with regular wind, but the turbines do not function in storms. If storms become frequent, the wind turbines will break. Then, the energy sources will be... Well, right now, they are strangling the energy sources, and some people are taking advantage of this situation. In other words, the world is currently moving toward abolishing the use of energies such as petroleum, coal, natural gas, and the use of power plants by 2050. This is actually what the evil space people are aiming for. If these energy sources cannot be used on Earth, human civilization would have to regress, which would consequently weaken the earthlings. Their combat capability will definitely drop, and their military power will weaken, too. That is what the evil space people are aiming for.

I also think that the nuclear disarmament movement is happening on Earth now because nuclear weapons are also troublesome for space people. There have been wars that resemble nuclear wars in outer space, too. There have been instances in which neither side was able to take the planet they were fighting for, and they both ended up having to abandon the planet to protect themselves. In that sense, space people also want Earth to be free of nuclear weapons.

In short, there are forces that want earthlings to regress toward a more primitive state. If you don't want that, I guess earthlings might have to escape to outer space.

A

In the dream, earthlings from 2100 were looking at a fossilized organism, right?

EDGAR CAYCE

No, I think it was about 2070.

A

About 2070. Then, what exactly does that have to do with the evil space people's plans? Is Earth habitable at that time? Are earthlings living on Earth?

EDGAR CAYCE

I don't know where the earthlings in spacesuits came from.

A

Oh, so do you mean it's not clear whether those earthlings are from Earth or somewhere else?

EDGAR CAYCE
Right. I don't know where they are from.

A
These people are from 2070, but they are still at a level of not knowing what it (the space slug) is.

EDGAR CAYCE
That's right.

A
They aren't able to figure out what it is yet.

EDGAR CAYCE
This also means there are others who know about a new energy source that is not found on Earth. This may be a discovery of a new source of wealth for them.

A
But the evil space people are currently using those organisms to collect the energy source to invade Earth.

EDGAR CAYCE
In Earth time, they spend hundreds of years coming to Earth, so they need enough power to continue traveling for that

long. There aren't many other planets like Earth in outer space.

I suspect that they will first send cyborgs to Earth before they themselves arrive on Earth in the next phase by dwelling in a life form that has been altered to suit life on Earth. At that time, they will likely try to eliminate any resistance they face. This is what will happen in the 21st century.

A

I see. That is frightening.

5
What the Potential Invaders of Earth Dislike the Most

Invaders of Earth vs. supporters of Earth

EDGAR CAYCE
Right now, you all are trying to keep a record of the Laws of the Universe and UFO readings. The future crises are really approaching, and there actually are space beings who are aiming to invade Earth. I'm not joking.

A
You mean there really are space people who are collecting energy by using a space sea cucumber or a space slug to maintain their UFOs near Earth and are planning to invade Earth?

EDGAR CAYCE
Yes. What's more, I believe the space people who are on your side have yet to talk to you about the energy source they are using.

A

You're right.

EDGAR CAYCE

They should also be using some sort of energy source.

At any rate, I sense that this will be a major space battle since the battle on Planet Beta, otherwise known as Planet Zeta. That is also why Ame-no-Mioya-Gami has appeared now. He is the main force of the Andromeda Galaxy.

A

Does that mean Ame-no-Mioya-Gami is here to help us?

EDGAR CAYCE

Yes, that's right. He is bringing reinforcements.

A

He really is.

EDGAR CAYCE

I think space people from the Andromeda Galaxy are strengthening their forces now. One of the origins of earthlings lies in the Andromeda Galaxy, so seeing that a crisis is approaching Earth, they are here to help you... They are increasing their forces now.

A

I see. The scale of this conversation is getting bigger and bigger. I've previously heard from Shakyamuni Buddha that Ame-no-Mioya-Gami descended on Andromeda to suppress the Reptilians or other space people who were trying to invade Earth.

EDGAR CAYCE

Yes.

A

So, that might mean that Earth will be the next battleground.

Carbon footprint reduction and nuclear disarmament movements are the enemy's trap

EDGAR CAYCE

Well, you would need an "outer space map." Without it, you won't be able to understand the various battles over territory that are happening across the entire universe. But these ideas are probably difficult for earthlings to understand based on their current level. The fact is, different beings have different thoughts and gods, so they all want to lead the universe in different directions.

Right now, certain earthlings are trying to control Earth by spreading ideas of stopping global warming, reducing CO_2 emissions, and ridding nuclear weapons. However, if earthlings are not careful, there is a chance of them falling into the enemy's hands by following these ideas. In that case, it may become difficult to defend Earth. It appears as though evil space people are trying to eliminate every single defensive function of Earth.

A
Well, I've heard that there is even a Chinese organization backing Ms. Greta.

EDGAR CAYCE
Yes. Yes. That's right. But China has no intention of reducing their CO_2 emissions or getting rid of their nuclear weapons.

A
Even the Black Lives Matter movement is apparently backed by a Chinese organization. They're everywhere. It's incredible.

EDGAR CAYCE
This level of greed for conquest is abnormal for an earthling. It's beyond the level of greed that earthlings typically have.

A

That's true. It's an incredible amount of greed.

What was brought in from space to destroy earthlings' faith in God

EDGAR CAYCE

You may not know this history, but when the Yuan Dynasty (the Mongol Empire) invaded Europe, they brought in a ton of bacteria that caused the plague. I doubt that earthlings at the time could develop any biological weapons, so there is a possibility that the bacteria were brought in from space.

A

I see. It was said that the plague came with the Mongols, right?

EDGAR CAYCE

That's right. The plague spread across Europe. It was one of the things that made European people suffer continuously from the Middle Ages to the modern day. They suffered significantly, and this is what destroyed their faith in God. The plague was brought in to destroy faith in God. It was as if the evil space people were mocking the earthlings by saying, "See if you can stop this. You can't stop this by praying." This is how evil space people tried to destroy people's faith. In that sense, it was a complicated battle.

A

Black Death or the plague was widespread at a time when humanity had become materialistic and had shallow faith to begin with...

EDGAR CAYCE

That's right.

A

So, it means that humanity is targeted by evil beings when they have shallow faith. What's more, the God of Earth does not approve of human beings who live like that, either...

EDGAR CAYCE

Right.

A

I wonder if the God of Earth saw the need to make humanity reflect on themselves...

EDGAR CAYCE

Well, for one thing, Black Death spread widely during the Crusades as well. This shows that perhaps some situations overlap between the times of the Crusades and the Yuan Dynasty. The other disease that has existed since a long time ago is smallpox, which may or may not have emerged naturally. It is possible for bacteria in the earthly world to naturally turn malignant, but there are also cases in which they turn malignant as a result of something evil dwelling in them.

Ultimately, it is easiest for evil to infiltrate when faith in the God of Earth is wavering.

In truth, there are many things in this world with the potential to turn malignant. They won't pose a problem when they are in harmony, but they turn malignant when harmony is disrupted.

Furthermore, I think evil space people will try to gradually control Earth's climate. For example, if something big like a hurricane hits Japan instead of the U.S., Japan in its current state would not stand a chance. Japan is used to dealing with typhoons, but it cannot handle a hurricane.

A

That's true.

EDGAR CAYCE
You might see unusual weather in the future.

A
Does that mean natural disasters that have never happened in Japan until now may happen in the future?

EDGAR CAYCE
Yes, that's right.

A
That's scary.

Systems that only exist on Earth

EDGAR CAYCE
So, you must be truly smart and wise to survive. You may not be able to survive with just ordinary tactics. There is no question that the enemy side wants to take control of Earth. I think they are hoping to create a new civilization of their own by using the life cycle system and the system for prosperity that only exist on Earth.

A

Are you referring to the system of reincarnation that the God of Earth has created specifically for Earth?

EDGAR CAYCE

Yes. They also want to change other things like the mechanism that allows humans to live on Earth.

A

Are they thinking of making their own version of this system to make something new, like their own planet?

EDGAR CAYCE

Yes, they are, especially because some of their own planets were destroyed. They probably think that they could take over the world if they could bring ruin to the U.S. and Europe as well as stop the development of their future enemies, like India and Brazil. Well... Living a long life is good, but those who do may experience some incredibly tough times...

A

If the God of Earth were not with us at that time, it would be dangerous, wouldn't it?

EDGAR CAYCE

Yes. I think it would be rather tough.

A

The evil space people like to use cyborgs, so does that mean earthlings will eventually be turned into cyborgs, too?

EDGAR CAYCE

Yes.

A

Based on that thinking, everyone except the enemy leader or "highest evil god" is bound to be degraded into a cyborg, no matter how I look at it. I think this will happen to humans.

EDGAR CAYCE

That is why they send in cyborgs that don't die from a viral infection.

A

Oh, I see. That's why.

EDGAR CAYCE

For example, cyborgs wouldn't die even if they are sent amidst the virus attack you are currently under.

A

The cyborgs would be the only ones to survive. How could that be?

EDGAR CAYCE

The cyborgs can keep fighting, but humans would die.

A

Humans wouldn't stand a chance.

EDGAR CAYCE

They would all die. So, evil space people are much further ahead in virus research. They are likely aiming to develop something that would only kill a particular species. This is not limited to humans; it also applies to animals. They would be in trouble if all the animals died, so they are trying to be selective about who and what to keep alive.

Well, if earthlings lose, they may have to escape from Earth to space like in the story of Noah's Ark. This is one possible scenario. That being said, even if humans escape to outer space, evil space people will pursue and continue fighting them because they can already travel through space.

A
How can we fight against such enemies?

EDGAR CAYCE
It is difficult because Japan is quite behind in the field of space.

A
You're right, we're significantly behind.

EDGAR CAYCE
Japan is lagging. They wasted their money on useless things.

A great shift is about to occur

EDGAR CAYCE
However, I think a great shift is about to occur. So, please look after HSU. A path to the future will open from there. As for the political party (Happiness Realization Party), I think it could take a different form and become the foundation that will lead the world based on its political opinion. That being said, there is no question that religion is what the enemies hate the most.

A

I wonder why that is. What is it about religion that they can't beat?

EDGAR CAYCE

On a fundamental level, they are totalitarian. The only thing that can beat totalitarianism is religion because even a single religion has the power to prevail on and protect the entire world.

A

It's also true that humans can demonstrate greater strength when they have faith.

EDGAR CAYCE

Yes, that's right.

A

They become more powerful.

EDGAR CAYCE

Humans can gain an even stronger other-dimensional power when they have faith. They won't be able to unleash their true powers as long as they think of themselves as a third-dimensional existence.

A

They will end up succumbing to scientific technology.

EDGAR CAYCE

Yes, exactly. This may go against your views, but if I were alive right now...

A

Are you going to say that you would choose to die early?

EDGAR CAYCE

... I sense that things could get quite dangerous on Earth from now on, so it's better to return to the other world soon.

A

Well, I agree with you on that. I think people who have already passed away are happier in a sense.

EDGAR CAYCE

You couldn't stand it if the movie *War of the Worlds*, which starred Tom Cruise, happened in reality, could you? Well, the scenario depicted in that movie is one possibility. But there are space people who are on your side, so you won't be attacked one-sidedly. In the movie, the invaders of Earth are defeated by viruses, but in Earth's current situation, the viruses are being used to wipe out humans. So, the strategy is being used against earthlings instead.

A
I see.

How to prevent space people from "walking in"

EDGAR CAYCE
This is a rather deep topic of discussion, so I alone cannot give a complete answer. All kinds of things which have yet to be revealed to you are involved, so I don't know the extent to which I am allowed to disclose information. I'm not sure whether humans will believe it and take action based on it.

A
Well, it all comes down to... People who have gathered in the Happy Science Sangha should not only be thinking about their own promotion or recognition to begin with. They must not hold Master back with their worldly attachments. Otherwise, he will not be able to preach what he needs to teach.

EDGAR CAYCE

At the very least, the trilateral alliance between Japan, the U.S., and Europe must be maintained. They need to create a system in which India and Brazil can develop into powerful countries, so that they can cooperate, too. I also think it is important to protect Australia from being taken over by China. Perhaps Australia was once destroyed in a past civilization. Given the large size of its desert region, it is likely that a civilization was once destroyed there.

Well, if I reveal the history of the rise and fall of the universe, it will start sounding like science fiction, so it will be very difficult to convince people to believe it. But I think the story should be told to some extent rather than having no story at all. It may become useful at some point.

A

I understand.

EDGAR CAYCE

You are at a very critical time now. Master is not making movies merely for entertainment purposes. He is offering the fundamental knowledge of space through the movies.

A
I agree wholeheartedly.

EDGAR CAYCE
He is teaching that there are various forces gathering in space now, and that there is a possibility of a fierce battle of invasion over Earth breaking out.

A
For us believers to reach that level of understanding, each of us is now being taught to stop being selfish and abandon our false selves.

EDGAR CAYCE
That is because they can be possessed...

A
Right, right. It's easy for evil beings to possess people who are selfish.

EDGAR CAYCE
It is to prevent evil space people from walking in...

A
That is the purpose of the teachings, isn't it?

EDGAR CAYCE

Yes, it is.

A

Master's teachings are related to everything, including the universe. It's about time the believers thought more about other people rather than thinking only about their own concerns. They have to learn to live in this way.

EDGAR CAYCE

If I were to add, I must say that Earth has been attacked many times by extraterrestrial forces in the over 300-million-year history of humankind. This has not been explained in detail, but there was a battle every time Earth was attacked, which is why God's teachings were given. In other words, those teachings were taught so that the invaders could not stay on Earth. They don't like those teachings.

A

They dislike them.

EDGAR CAYCE

To evil space people, these teachings weaken their power. They have the same effect that naphthalene, blue mold, or penicillin have on germs.

A

That is why some people think their own power gets weakened when they listen to Shakyamuni Buddha's teachings...

EDGAR CAYCE

Their power gets weakened...

A

It means they have become people who are susceptible to virus attacks, so they need to make an effort to change their minds.

EDGAR CAYCE

Right, right, right.

A

But the truth is that you *don't* get weaker.

EDGAR CAYCE

If more and more people become selfish, in other words, if more and more people become materialistic and only go after worldly success...

A

They will be the same as the enemies.

EDGAR CAYCE
It will become easier for space people to invade Earth...

A
These people will come to have the same mindset as the invaders.

EDGAR CAYCE
Humans will become like the invaders. Well, I think these ideas need to be studied more. This means that being attached to this world draws the invaders closer and closer to you.

A
That's true. I see.

What humans should do before the Laws of the Universe are taught

EDGAR CAYCE
Well, this has become a much bigger discussion than expected.

A
That's OK. In the recent issues of *Are You Happy?* magazine and the Happy Science monthly magazine, there were stories

of Happy Science believers who have been spreading Master's teachings to others for decades. Their stories have encouraged me a lot. You are saying that producing more people like them is the best way to win against evil forces, right?

EDGAR CAYCE

Well, the Laws of the Universe is the last teaching that will be preached, so in preparation for that, Happy Science must increase its number of believers. If there are only a few believers, the teachings will simply look heretical. If other people start attacking the teachings as heresy, they will start to join hands with the evil forces that want to destroy the teachings. For this reason, believers must change the minds of people before then. Well, it is difficult to do, indeed.

A

Today's discussion has really become a much bigger discussion than expected.

EDGAR CAYCE

I guess this dream interpretation turned out to be that way. Well, you might be able to use it in your next movie series (*The Laws of the Universe*).

A

Yes, that's true. *The Age of El Cantare* is anticipated to be the third sequel (in *The Laws of the Universe* movie series).

EDGAR CAYCE

That's right. (At the time of the recording. The movie, *The Laws of the Universe—The Age of El Cantare—* has been canceled at this point in time.)

A

Thank you very much.

EDGAR CAYCE

Well, you all have a tough mission ahead of you. You may end up in a situation where you can neither live nor die, even if you want to.

A

Right.

EDGAR CAYCE

Yes.

RYUHO OKAWA

OK, then [*claps once*].

A

Thank you very much.

RYUHO OKAWA

Sure.

For a deeper understanding of
Ryuho Okawa's Spiritual Trip to Mars
see other books below by Ryuho Okawa:

The Laws of Eternity [New York: IRH Press, 2024]

The Xi Jinping Thought Now [Tokyo: HS Press, 2021]

R. A. Goal's Words for the Future [Tokyo: HS Press, 2021]

The Descent of Japanese Father God Ame-no-Mioya-Gami [Tokyo: HS Press, 2022]

ABOUT THE AUTHOR

Founder and CEO of Happy Science Group.

Ryuho Okawa was born on July 7th, 1956, in Tokushima, Japan. After graduating from the University of Tokyo with a law degree, he joined a Tokyo-based trading company. While working at its New York headquarters, he studied international finance at the Graduate Center of the City University of New York. In 1981, he attained Great Enlightenment and became aware that he is El Cantare with a mission to bring salvation to all humankind.

In 1986, he established Happy Science. It now has members in 188 countries across the world, with more than 700 branches and temples, as well as 10,000 missionary houses around the world.

He has given over 3,500 lectures (of which more than 150 are in English) and published over 3,300 books (of which more than 600 are Spiritual Interview Series), and many are translated into 42 languages. Along with *The Laws of the Sun* and *The Laws of Hell*, many of the books have become best sellers or million sellers. To date, Happy Science has produced 29 movies under his supervision. He has given the original story and concept and is also the Executive Producer. He has also composed music and written lyrics for over 450 pieces.

Moreover, he is the Founder of Happy Science University and Happy Science Academy (Junior and Senior High School), Founder and President of the Happiness Realization Party, Founder and Honorary Headmaster of Happy Science Institute of Government and Management, Founder of IRH Press Co., Ltd., and the Chairperson of NEW STAR PRODUCTION Co., Ltd. and ARI Production Co., Ltd.

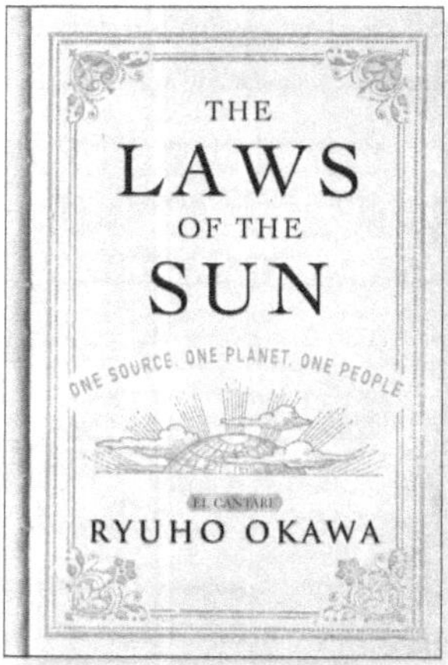

THE LAWS OF THE SUN

ONE SOURCE, ONE PLANET, ONE PEOPLE

ISBN: 978-1-942125-43-3 • US$15.95

Why and how did God create this world? What is a spirit and soul? For what purpose are humans living in this world? How can we find true happiness and meaning in life?

The Laws of the Sun answers life's questions humans have always had throughout history.

In this book, Ryuho Okawa outlines the laws that govern the universe and provides a road map for living one's life with greater purpose and meaning.

The Truth taught in this book will significantly transform your life and allow you to develop love and acceptance toward people of all races and religions.

2026 New Titles

THE CHALLENGE TO ESTABLISHING THE IMPERISHABLE TRUTH

NOW IS THE AGE OF SPIRITUAL AWAKENING

ISBN: 978-1958655399 • US$17.95

This book encourages readers to establish a spiritual view of life, understanding that life in this world is a soul-training ground for deepening love and enlightenment.

INVITATION TO FAITH

ATTAIN ENLIGHTENMENT AND GIVE LOVE

ISBN: 978-1-958655-71-9 • US$17.95

This world is not the only world there is, and life continues after death. The problems we face in life are actually chances for us to attain wisdom and cultivate selfless, pure love for others. This book presents the universal faith that people of today are seeking deep inside.

FROM LOVE TO PRAYER

THE LIGHT OF HERMES DESCENDS ONCE AGAIN

ISBN: 978-1958655412 • US$17.95

This is an enlightening collection of poetic verses in the form of prayers that gently guides readers toward inner peace, spiritual awakening, and a deeper understanding of the true nature of love and prayer. Through these noble passages, readers will discover the truth of their souls.

Books for the Coming Space Age

THE LAWS OF BRONZE

LOVE ONE ANOTHER, BECOME ONE PEOPLE

ISBN: 978-1-942125-50-1 • US$15.95

We have now entered a new space age. Chapter 4 reveals the involvement of space people in the history of humankind and Happy Science's mission to create a common ground to unite Earth as one.

KANJIZAI, BUDDHA'S OMNISCIENT POWER OF PERCEPTION

GOING BEYOND TIME, SPACE, AND THE GREAT UNIVERSE

ISBN: 978-1958655290 • US$17.95

This book unveils what it means to possess real spiritual power–one that inspires humility, sincerity, and the pursuit of genuine enlightenment.

THE TRUTH ABOUT EARTH, THE UNIVERSE, THE SPIRIT WORLD

LIFE'S Q&A WITH EL CANTARE

ISBN: 978-1-958655-26-9 • US$17.95

A compilation of 28 Q&A sessions by Ryuho Okawa, where he answers numerous intricate questions varying from the theory of evolution to the creation of the multidimensional universe, all without a script.

UFO READING LIGHT VS. DARKNESS

THE CLASHING VALUES OF THE UNIVERSE

ISBN: 978-4-8233-0454-5 • US$15.95

In these captivating telepathic exchanges between the author Ryuho Okawa and various extraterrestrial beings, you will learn that a fierce battle over the values of good and evil has been taking place between alien allies and adversaries.

WITH SAVIOR

MESSAGES FROM SPACE BEING YAIDRON

ISBN: 978-1-943928-17-0 • US$13.95

The human race is now faced with multiple crises. Perhaps God is warning us to reconsider our materialistic and arrogant ways. Fortunately, God has sent us a savior, who is now showing us the path we should choose. In this book, space being Yaidron sends his warnings and messages of hope.

SPIRITUAL MESSAGES FROM YAIDRON

SAVE THE WORLD FROM DESTRUCTION

ISBN: 979-8-88737-001-9 • US$11.95

Yaidron is a certified messiah from Planet Elder in the Magellanic Clouds. In this book, he explains what was occurring behind the coup in Myanmar and Taliban's control over Afghanistan, and warns of the imminent danger approaching Taiwan.

SPIRITUAL MESSAGES FROM METATRON

LIGHT IN THE TIMES OF CRISIS

ISBN: 978-1-943928-19-4 • US$11.95

Metatron, a certified messiah from Planet Include in Sagittarius, has sent a message upon seeing the crisis of Earth. You will learn the real intention of China and the danger of Japan's appeasement policy toward China.

R. A. GOAL'S WORDS FOR THE FUTURE

MESSAGES FROM A SPACE BEING TO THE PEOPLE OF EARTH

ISBN: 979-8-887370-52-1 • US$11.95

R. A. Goal is a certified messiah from Planet Andalucia Beta in Ursa Minor, who protects and guides us. By reading this book, you will have a clear picture of how the world now needs true leaders who can think about world justice.

Spiritual Truth

THE LAWS OF ETERNITY

EL CANTARE UNVEILS THE STRUCTURE OF THE SPIRIT WORLD

ISBN: 978-1-958655-16-0 • US$17.95

"Where do we come from and where do we go after death?" *The Laws of Eternity* answers life's essential questions we are all confronted with at some point. This book takes you on a journey to the other world, a place where humans came from and return to after death.

THE LAWS OF MESSIAH

FROM LOVE TO LOVE

ISBN: 978-1-942125-90-7 • US$16.95

"What is Messiah?" This book carries an important message of love and guidance to people living now from the Modern-Day Messiah or the Modern-Day Savior. It also reveals the secret of Shambhala, the spiritual center of Earth.

THE LAWS OF HELL

"IT" FOLLOWS.....

ISBN: 978-1-958655-04-7 • US$17.95

Whether you believe it or not, the Spirit World and hell do exist, and unfortunately, 1 in 2 people are falling into hell. To stop hell from spreading and to save the souls of all human beings, Ryuho Okawa has compiled vital teachings in this book.

THE MYSTICAL LAWS

GOING BEYOND THE DIMENSIONAL BOUNDARIES

ISBN: 979-8-887370-3-85 • US$14.95

"Within an hour of reading this book, all that you have accepted as common knowledge will crumble away. Indeed, no school or mass media has ever taught you the truth. Whatever doubts you may have, this book will reveal the truths of the great universe." —From Preface

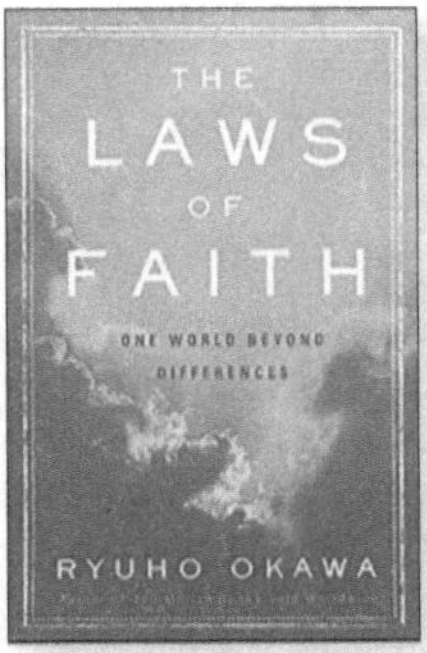

THE LAWS OF FAITH

ONE WORLD BEYOND DIFFERENCES

ISBN: 978-1-942125-34-1 • US$15.95

Ryuho Okawa preaches the core teachings of the world religion and the faith in the God of Earth. Through this book, you will learn to go beyond different values, harmonize with each other and between nations, and create a world filled with peace and prosperity.

SECRETS OF THE EVERLASTING TRUTHS

A NEW PARADIGM FOR LIVING ON EARTH

ISBN: 978-1-937673-10-9 • US$14.95

Ryuho Okawa offers a glimpse of the vast universe created by God and discloses that humanity is intimately guided by celestial influences. Our planet will experience a decisive paradigm shift of "knowledge" and "truth" as the new era unfolds.

Love

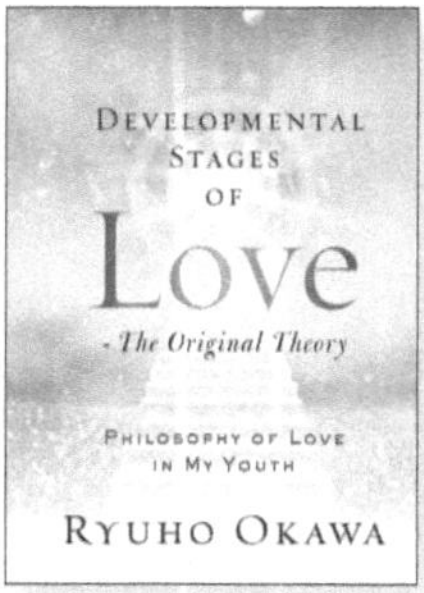

DEVELOPMENTAL STAGES OF LOVE —THE ORIGINAL THEORY

PHILOSOPHY OF LOVE IN MY YOUTH

ISBN: 978-1-942125-94-5 • US$17.95

A series of short essays on love written by the author when he was 25 to 28. This is Ryuho Okawa's original theory on love before he became a religious leader. Open its pages and discover true love.

BE INFINITELY KIND

FOR WE ARE LIVING IN THE GREAT RIVER OF LOVE

ISBN: 978-1-958655276 • US$17.95

In this book, Ryuho Okawa invites readers to explore the deeper spiritual virtues of love and kindness—essential qualities for building strong, lasting relationships with those we hold dear. Truthfully, the more love we give, the wealthier our souls become.

THE TEN PRINCIPLES FROM EL CANTARE VOLUME I

RYUHO OKAWA'S FIRST LECTURES ON HIS BASIC TEACHINGS

ISBN: 978-1-942125-85-3 • US$16.95

How did Happy Science begin? What are its teachings? What is its aim? This is a compilation of early lectures that built the foundation of Happy Science. The historic moments are captured in this book.

THE TEN PRINCIPLES FROM EL CANTARE VOLUME II

RYUHO OKAWA'S FIRST LECTURES ON HIS WISH TO SAVE THE WORLD

ISBN: 978-1-942125-86-0 • US$16.95

Volume II contains passionate messages of Ryuho Okawa filled with the wish to save humankind and build utopia on earth. You can also learn about God's three major inventions, the mission of religion, and more.

WHAT IS HAPPY SCIENCE?

BEST SELECTION OF RYUHO OKAWA'S EARLY LECTURES (VOLUME 1)

ISBN: 978-1-942125-99-0 • US$17.95

The Best Selection series is a collection of Ryuho Okawa's lectures given when he was 32 to 33 that reveal the mission and goal of Happy Science. Volume 1 teaches the eternal Truth, including the secret of the mind, the true meaning of love, and how to end hatred and world conflicts.

THE WAY TO HUMAN PERFECTION

BEST SELECTION OF RYUHO OKAWA'S EARLY LECTURES (VOLUME 2)

ISBN: 978-1-958655-20-7 • US$17.95

The path to enlightenment starts from understanding "the eternal viewpoint of life." By recognizing that we have eternal life, we can realize that caring and bringing joy to others are the keys to true happiness and success.

Happiness

THE STARTING POINT OF HAPPINESS

AN INSPIRING GUIDE TO POSITIVE LIVING WITH FAITH, LOVE, AND COURAGE

ISBN: 978-1-942125-26-6 • US$16.95

This book is for those who are struggling to attain love and happiness. It reveals the simple mindsets that will help you overcome difficult circumstances and become happier.

THE LAWS OF HAPPINESS

LOVE, WISDOM, SELF-REFLECTION AND PROGRESS

ISBN: 978-1-942125-70-9 • US$16.95

Happiness is found within us, not outside. It is in how we think, view life, and work. This is an ultimate guide to becoming happy through the Fourfold Path of Love, Wisdom, Self-Reflection, and Progress—the universal law of happiness.

THE UNHAPPINESS SYNDROME

28 HABITS OF UNHAPPY PEOPLE (AND HOW TO CHANGE THEM)

ISBN: 978-1-942125-16-7 • US$15.95

Have you fallen into a cycle of pessimistic attitudes? You may have the unhappiness syndrome, which is a subconscious love for unhappiness. Ryuho Okawa identifies the 28 common habits of unhappy people and provides remedies to counter them.

WORRY-FREE LIVING

LET GO OF STRESS AND LIVE IN PEACE AND HAPPINESS

ISBN: 978-1-942125-51-8 • US$16.95

This is a guide for those who are overwhelmed by the stress that comes with living. It will help you change how you look at and approach life's worries and problems for the better.

AN UNSHAKABLE MIND

HOW TO OVERCOME LIFE'S DIFFICULTIES

ISBN: 978-1-942125-91-4 • US$17.95

This book will guide you to build the genuine self-confidence necessary to shape a resilient character and withstand life's turbulence. Ryuho Okawa breaks down the causes of life's difficulties and provides solutions to overcome them.

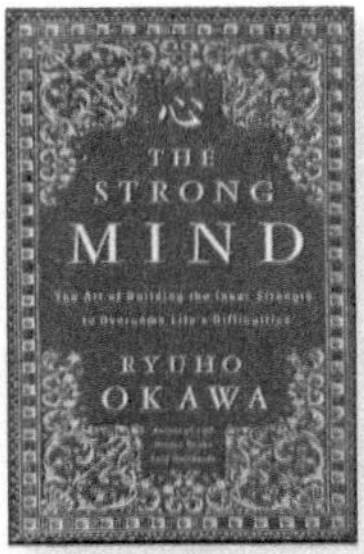

THE STRONG MIND

THE ART OF BUILDING THE INNER STRENGTH TO OVERCOME LIFE'S DIFFICULTIES

ISBN: 978-1-942125-36-5 • US$15.95

This book will inspire and empower you to take courage, develop a cultivated heart, and achieve resilience and hardiness so that you can break through your limits and keep winning in the battle of your life.

THE ROYAL ROAD OF LIFE

BEGINNING YOUR PATH OF INNER PEACE, VIRTUE, AND A LIFE OF PURPOSE

ISBN: 978-1-942125-53-2 • US$16.95

What is the essence of a virtuous leader? *The Royal Road of Life* introduces a way of life based on understanding the nature of your mind and gaining mastery over it; it is a guide to maintaining peace of mind and cultivating virtue throughout your life.

Success

THE LAWS OF SUCCESS

A SPIRITUAL GUIDE TO TURNING YOUR HOPES INTO REALITY

ISBN: 978-1-942125-15-0 • US$15.95

The Laws of Success offers 8 spiritual principles that will help us attain lasting success. The timeless wisdom and practical steps that Ryuho Okawa offers will serve as guiding principles for living a positive, constructive, and meaningful life.

THE ART OF INFLUENCE

28 WAYS TO WIN PEOPLE'S HEARTS AND BRING POSITIVE CHANGE TO YOUR LIFE

ISBN: 978-1-942125-48-8 • US$15.95

This is a compilation of 28 Q&A sessions. Inside are Okawa's deep wisdom that will offer you motivation and inspiration, and develop you into an influential leader with a big heart.

THINK BIG!

BE POSITIVE AND BE BRAVE TO ACHIEVE YOUR DREAMS

ISBN: 978-1-942125-04-4 • US$12.95

Think Big! offers the support and encouragement to shift to new ways of thinking and mastering self-discipline. The self-proven approach fosters stability and strength in the challenges we face.

THE HEART OF WORK

10 KEYS TO LIVING YOUR CALLING

ISBN: 978-1-942125-03-7 • US$12.95

In this book, Ryuho Okawa introduces 10 key philosophies to live by to guide you through your work life. By grasping them, you can live your calling with utmost confidence and an unshakable mind.

THE LAWS OF HOPE

THE LIGHT IS HERE

ISBN:978-1-942125-76-1 • US$16.95

This book provides ways to bring light and hope to ourselves, even in the midst of sufferings and adversities. By making efforts in your current circumstances, you can fulfill your mission to shed light on yourself and those around you.

THE POWER OF BASICS

INTRODUCTION TO THE MODERN ZEN LIFE OF CALM, SPIRITUALITY AND SUCCESS

ISBN:978-1-942125-75-4 • US$16.95

The power of basics is a necessary asset to excel at any kind of work. It is the power to meticulously pursue tasks with a quiet Zen mindset. By mastering it, you can attain new heights of success and happiness.

INVINCIBLE THINKING

AN ESSENTIAL GUIDE FOR A LIFETIME OF GROWTH, SUCCESS, AND TRIUMPH

ISBN: 978-1-942125-25-9 • US$16.95

This book introduces the principles of invincible thinking that will allow us to achieve long-lasting triumph. By learning from every experience and transforming it into wisdom, we will never be defeated.

THE LAWS OF INVINCIBLE LEADERSHIP

AN EMPOWERING GUIDE FOR CONTINUOUS AND LASTING SUCCESS IN BUSINESS AND IN LIFE

ISBN: 978-1-942125-30-3 • US$19.95

Ryuho Okawa shares essential principles for all who wish to become invincible managers and leaders in their fields of work, organizations, societies, and nations.

Health & Wellness

HEALING FROM WITHIN

LIFE-CHANGING KEYS TO CALM, SPIRITUAL, AND HEALTHY LIVING

ISBN:978-1-942125-18-1 • US$15.95

In this book, Ryuho Okawa reveals the true causes and remedies for various illnesses that modern medicine cannot heal. You will learn about the relationship between mind and body.

THE MIRACLE POWER TO OVERCOME ILLNESS

HEALING THROUGH FAITH

ISBN: 978-1-958655-17-7 • US$17.95

The truth inside this book will take you beyond the worldly understanding of illness and help you awaken to the power of faith. Learn to become a person who can receive the miracle power from heaven.

WORDS TO READ IN TIMES OF ILLNESS

ISBN: 978-1-958655-07-8 • US$17.95

Ryuho Okawa has written 100 Healing Messages to comfort the souls of those going through illness. A time of illness is an ideal time for us to contemplate recent and past events, as well as our relationship with the people around us.

THE MIRACLE OF MEDITATION

OPENING YOUR LIFE TO PEACE, JOY, AND THE POWER WITHIN

ISBN: 978-1-942125-09-9 • US$15.95

This book is your entrance to understanding the real meaning of meditation and how to correctly meditate for your self-growth. By combining faith with meditation, you can discover your inner divinity.

Enlightenment

THE ESSENCE OF BUDDHA

THE PATH TO ENLIGHTENMENT

ISBN: 978-1-942125-06-8 • US$14.95

The essence of Shakyamuni Buddha's original teachings of the mind is explained in simple words. Through this book, you will learn how to attain inner happiness, the wisdom to conquer ego, and to enter the path to enlightenment.

THE LAWS OF GREAT ENLIGHTENMENT

ALWAYS WALK WITH BUDDHA

ISBN: 978-1-942125-62-4 • US$17.95

Discover the power of forgiveness from Buddha's enlightenment and compassion, and the true relationship of work and enlightenment. In addition, the author reveals his own experience when he attained Great Enlightenment.

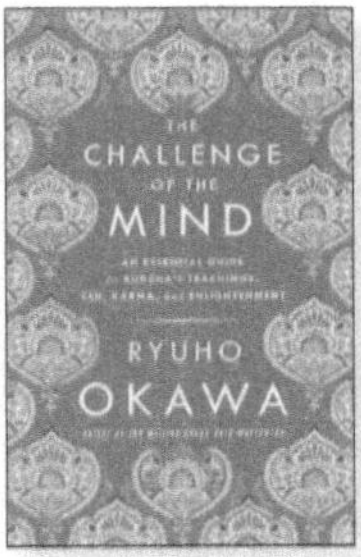

THE CHALLENGE OF THE MIND

AN ESSENTIAL GUIDE TO BUDDHA'S TEACHINGS: ZEN, KARMA, AND ENLIGHTENMENT

ISBN: 978-1-942125-45-7 • US$16.95

In this book, Ryuho Okawa explains essential Buddhist tenets and how to put them into practice. He offers a solid basis of reason and an intellectual understanding of Buddhist concepts.

THE TRUE EIGHTFOLD PATH

GUIDEPOSTS FOR SELF-INNOVATION

ISBN: 978-1-942125-80-8 • US$16.95

This book explains how we can apply the Eightfold Path, one of the main pillars of Shakyamuni Buddha's teachings, as everyday guideposts in the modern age to achieve self-innovation to live better and make positive changes.

Enlightenment

The Challenge of Enlightenment

Now, Here, the New Dharma Wheel Turns

ISBN: 978-1-942125-92-1 • US$17.95

What is the role of enlightenment in facing the challenges of the modern world? Find out how to savor inner peace, and be empowered to make profound life improvements.

The Rebirth of Buddha

My Eternal Disciples, Hear My Words

ISBN: 978-1-942125-95-2 • US$17.95

These are the messages of Buddha, who has returned to this modern age as promised to his eternal beloved disciples. They are in simple words and poetic style, yet contain profound messages.

The Eternal Buddha

Now, Here, Is the Imperishable Light

ISBN: 978-1-958655-19-1 • US$17.95

Embedded in this book is a message from Eternal Buddha, the parent of your soul. You will discover the true origin of your soul and why you have chosen to be born in this time.

Rojin, Buddha's Mystical Power

Its Ultimate Attainment in Today's World

ISBN: 978-1-942125-82-2 • US$16.95

In this book, Ryuho Okawa has redefined the traditional Buddhist term Rojin in the context of modern society. This book will unravel the mystery of the mind and lead you to the path to enlightenment.

The Descent of Japanese Father God Ame-no-Mioya-Gami

The "God of Creation" in the Ancient Document *Hotsuma Tsutae*

ISBN: 978-1-943928-35-4 • US$14.95

Nearly 30,000 years ago, Ame-no-Mioya-Gami descended on the foothills of Mt. Fuji. By reading this book, you can find the origin of the Japanese spirit and understand how the ancient Japanese civilization influenced other countries.

True Bushido Spoken By Ame-no-Mioya-Gami

The Japanese Father God Teaches How We Should Live and Die

ISBN: 978-4823304606 • US$20.00

In this book, Ame-no-Mioya-Gami speaks to the people living in modern society in the form of a spiritual message. Read and discover the true meaning behind bushido and why people must regain the samurai spirit.

WHO IS EL CANTARE?

El Cantare means "the Light of the Earth." He is the Supreme God of the Earth who has been guiding humankind since the beginning of Genesis, and He is the Creator of the universe. He is whom Jesus called Father and Muhammad called Allah, and He is *Ame-no-Mioya-Gami*, the Japanese Father God. He is also known as Vishnu in India and Tiandi in China. Different parts of El Cantare's core consciousness have descended to Earth in the past, once as Alpha and another as Elohim. His branch spirits, such as Shakyamuni Buddha and Hermes, have descended to Earth many times and helped to flourish many civilizations. To build a new civilization on Earth by uniting various religions and integrating various fields of study, a part of the core consciousness has now descended as Master Ryuho Okawa.

Alpha is a part of the core consciousness of El Cantare, who descended to Earth around 330 million years ago. Alpha preached Earth's Truth to harmonize and unify Earth-born humans and space people who came from other planets.

Elohim is a part of the core consciousness of El Cantare, who descended to Earth around 150 million years ago. He gave wisdom, mainly on the differences between light and darkness, good and evil.

Ame-no-Mioya-Gami (The Japanese Father God) is a Being who is close to the core consciousness of Lord El Cantare, and is the Creator God and the Father God who appears in ancient literature, *Hotsuma Tsutae*. It is believed that He descended on the foothills of Mt. Fuji about 30,000 years ago and built the Fuji dynasty, which is the root of the Japanese civilization. Ame-no-Mioya-Gami's Teachings spread to ancient civilizations of other countries in the world.

EL CANTARE'S BRANCH SPIRITS

Shakyamuni Buddha was born in Nepal around 2,600 years ago as the prince of the Shakya clan. When he was 29 years old, he renounced the world and sought enlightenment. He later attained Great Enlightenment and spent most of his life in India teaching and practicing the Truth. He is the founder of Buddhism, which has spread extensively throughout Asia.

Hermes is one of the 12 Olympian gods in Greek mythology, but the spiritual Truth is that he taught the teachings of love and progress around 4,300 years ago, which became the origin of the current Western civilization. He is a hero who truly existed.

Ophealis was born in Greece around 6,500 years ago and was the leader who went on an expedition as far as Egypt. He is the God of miracles, prosperity, and arts, and is known as Osiris in Egyptian mythology.

Rient Arl Croud was born as a king of the ancient Incan Empire around 7,000 years ago and taught about the mysteries of the mind. In the heavenly world, he is responsible for the interactions that take place between Earth and various planets.

Thoth was an almighty leader who built the golden age of the Atlantis civilization around 12,000 years ago. In Egyptian mythology, he is known as God Thoth.

Ra Mu was a leader who built the golden age of the civilization of Mu around 17,000 years ago. As a religious leader and a politician, he ruled by uniting religion and politics.

ABOUT HAPPY SCIENCE

Happy Science is a religious group founded on the faith in El Cantare, who is the God of the Earth and the Creator of the universe. The true essence of human beings is an eternal soul created by God, and we go through the cycle of reincarnation to train and develop our souls. We have been carrying out various activities to spread this spiritual value and build a peaceful and prosperous world that God wishes. At Happy Science, we explore righteousness, in other words, God's Will. This is called the "Exploration of Right Mind." More specifically, it means to practice the Fourfold Path: Love, Wisdom, Self-Reflection, and Progress. This is the path that leads humans to attain happiness that carries over from this world to the next.

Love–practicing "love that gives"
To give love to others without expecting anything in return is the starting point of happiness. By practicing "love that gives," you will become closer to God.

Wisdom—studying spiritual truth
By studying the Truth, you will be able to distinguish good and evil, live righteously, and learn the heart of God. True wisdom leads people to true happiness.

Self-Reflection—correcting your mistakes
Self-reflection is the act of correcting wrongful thoughts and actions you have accumulated during the day and regaining the pure, true nature of the soul.

Progress—creating utopia on earth
True progress is about spreading happiness to others and improving society as you achieve your own success. This will create utopia on earth.

PLACES OF WORSHIP FOR HAPPY SCIENCE

Shoja

—A PLACE TO REFINE YOUR MIND, GAIN SPIRITUAL WISDOM, AND BE REBORN

Happy Science shoja (temple) is a sacred spiritual field where you can deepen your faith and heighten your enlightenment. Under the spiritual guidance of high spirits, shoja holds various seminars to improve individual character and ritual prayers to help believers solve their life problems and make progress. By participating in them, you can regain the peaceful and blissful mind that you may have lost in your everyday life.

29 SHOJAS IN JAPAN, 3 SHOJAS AND LA DOJO OVERSEAS

San Francisco

Local Branch

—A PLACE TO CHANGE YOUR DESTINY

Since 1986, Happy Science has been carrying out various activities to produce people who can truly say, "I am happy." At our local branches, many new believers are being born and are leading better lives through faith in Lord God El Cantare. All kinds of events take place here, such as lecture viewing, book seminars, prayers, and counseling sessions. Everyone is welcome!

"The True Words Spoken By Buddha"

"The True Words Spoken By Buddha" is an English sutra given directly from the spirit of Shakyamuni Buddha, who is a part of Master Ryuho Okawa's subconscious. The words in this sutra are not of a mere human being but are the words of God or Buddha sent directly from the ninth dimension, which is the highest realm of the Earth's Spirit World.

"The True Words Spoken By Buddha" is an essential sutra for us to connect and live with God or Buddha's Will as our own.

MEMBERSHIPS

MEMBERSHIP

If you would like to know more about Happy Science, please consider becoming a member. Those who pledge to believe in Lord El Cantare and wish to learn more can join us.

When you become a member, you will receive the following sutras: "The True Words Spoken By Buddha," "Prayer to the Lord" and "Prayer to Guardian and Guiding Spirits."

DEVOTEE MEMBER

If you would like to learn the teachings of Happy Science and walk the path of faith, become a Devotee member who pledges devotion to the Three Treasures, which are Buddha, Dharma, and Sangha. Buddha refers to Lord El Cantare, Master Ryuho Okawa. Dharma refers to Master Ryuho Okawa's teachings. Sangha refers to Happy Science. Devoting to the Three Treasures will let your Buddha nature shine, and you will enter the path to attain true freedom of the mind.

Becoming a devotee means you become Buddha's disciple. You will discipline your mind and act to bring happiness to society.

EMAIL OR **PHONE CALL**

Please turn to the contact information page.

ONLINE member.happy-science.org/signup/

CONTACT INFORMATION

Happy Science is a worldwide organization with branches and temples around the globe. For full details, visit happy-science.org. The following are some of our main Happy Science locations:

UNITED STATES AND CANADA

New York
79 Franklin St., New York, NY 10013, USA
Phone: 1-212-343-7972
Fax: 1-212-343-7973
Email: ny@happy-science.org
Website: happyscience-usa.org

New Jersey
66 Hudson St., #2R, Hoboken, NJ 07030, USA
Phone: 1-201-313-0127
Email: nj@happy-science.org
Website: happyscience-usa.org

Chicago
33 West Higgins Rd. 4040,
South Barrington, IL 60010, USA
Phone: 1-630-937-3077
Email: chicago@happy-science.org
Website: happyscience-usa.org

Florida
5208 8th St., Zephyrhills, FL 33542, USA
Phone: 1-813-715-0000
Fax: 1-813-715-0010
Email: florida@happy-science.org
Website: happyscience-usa.org

Atlanta
1874 Piedmont Ave., NE Suite 360-C
Atlanta, GA 30324, USA
Phone: 1-404-892-7770
Email: atlanta@happy-science.org
Website: happyscience-usa.org

San Francisco
525 Clinton St.
Redwood City, CA 94062, USA
Phone & Fax: 1-650-363-2777
Email: sf@happy-science.org
Website: happyscience-usa.org

Los Angeles
1590 E. Del Mar Blvd., Pasadena,
CA 91106, USA
Phone: 1-626-395-7775
Fax: 1-626-395-7776
Email: la@happy-science.org
Website: happyscience-usa.org

Orange County
16541 Gothard St. Suite 104
Huntington Beach, CA 92647
Phone: 1-714-659-1501
Email: oc@happy-science.org
Website: happyscience-usa.org

San Diego
7841 Balboa Ave. Suite #202
San Diego, CA 92111, USA
Phone: 1-626-395-7775
Fax: 1-626-395-7776
E-mail: sandiego@happy-science.org
Website: happyscience-usa.org

Hawaii
Phone: 1-808-591-9772
Fax: 1-808-591-9776
Email: hi@happy-science.org
Website: happyscience-usa.org

Kauai
3343 Kanakolu Street, Suite 5
Lihue, HI 96766, USA
Phone: 1-808-822-7007
Fax: 1-808-822-6007
Email: kauai-hi@happy-science.org
Website: happyscience-usa.org

Toronto
845 The Queensway
Etobicoke, ON, M8Z 1N6, Canada
Phone: 1-416-901-3747
Email: toronto@happy-science.org
Website: happy-science.ca

Vancouver
#201-2607 East 49th Avenue,
Vancouver, BC, V5S 1J9, Canada
Phone: 1-604-437-7735
Fax: 1-604-437-7764
Email: vancouver@happy-science.org
Website: happy-science.ca

WORLDWIDE

Tokyo
1-6-7 Togoshi, Shinagawa,
Tokyo, 142-0041, Japan
Phone: 81-3-6384-5770
Fax: 81-3-6384-5776
Email: tokyo@happy-science.org
Website: happy-science.org

London
3 Margaret St.
London, W1W 8RE United Kingdom
Phone: 44-20-7323-9255
Fax: 44-20-7323-9344
Email: eu@happy-science.org
Website: www.happyscience-uk.org

Sydney
516 Pacific Highway, Lane Cove North,
2066 NSW, Australia
Phone: 61-2-9411-2877
Fax: 61-2-9411-2877
Email: aus@happy-science.org

Sao Paulo
Rua. Domingos de Morais 1154,
Vila Mariana, Sao Paulo SP
CEP 04010-100, Brazil
Phone: 55-11-5088-3800
Email: sp@happy-science.org
Website: happyscience.com.br

Jundiai
Rua Congo, 447, Jd. Bonfiglioli
Jundiai-CEP, 13207-340, Brazil
Phone: 55-11-4587-5952
Email: jundiai@happy-science.org

Seoul
74, Sadang-ro 27-gil,
Dongjak-gu, Seoul, Korea
Phone: 82-2-3478-8777
Fax: 82-2-3478-9777
Email: korea@happy-science.org

Taipei
No. 89, Lane 155, Dunhua N. Road,
Songshan District, Taipei City 105, Taiwan
Phone: 886-2-2719-9377
Fax: 886-2-2719-5570
Email: taiwan@happy-science.org

Taichung
No. 146, Minzu Rd., Central Dist.,
Taichung City 400001, Taiwan
Phone: 886-4-22233777
Email: taichung@happy-science.org

Kuala Lumpur
No 22A, Block 2, Jalil Link Jalan Jalil Jaya
2, Bukit Jalil 57000,
Kuala Lumpur, Malaysia
Phone: 60-3-8998-7877
Fax: 60-3-8998-7977
Email: malaysia@happy-science.org
Website: happyscience.org.my

Kathmandu
Kathmandu Metropolitan City,
Ward No. 15, Ring Road, Kimdol,
Sitapaila Kathmandu, Nepal
Phone: 977-1-537-2931
Email: nepal@happy-science.org

Kampala
Plot 877 Rubaga Road, Kampala
P.O. Box 34130 Kampala, Uganda
Email: uganda@happy-science.org

HAPPY SCIENCE UNIVERSITY

THE FOUNDING SPIRIT AND THE GOAL OF EDUCATION

Based on the founding philosophy of the university, "Exploration of happiness and the creation of a new civilization," education, research, and studies will be provided to help students acquire a deep understanding grounded in religious belief and advanced expertise with the objectives of producing "great talents of virtue" who can contribute in a broad-ranging way to serve Japan and the international society.

FACULTIES

Faculty of Human Happiness

Students in this faculty will pursue liberal arts from various perspectives with a multidisciplinary approach, explore and envision an ideal state of human beings and society.

Faculty of Successful Management

This faculty aims to realize successful management that helps organizations create value and wealth for society and to contribute to the happiness and development of management and employees, as well as society as a whole.

Faculty of Future Creation

Students in this faculty study subjects such as political science, journalism, performing arts, and artistic expression, and explore and present new political and cultural models based on truth, goodness, and beauty.

Faculty of Future Industry

This faculty aims to nurture engineers who can resolve various issues facing modern civilization from a technological standpoint and contribute to the creation of new industries of the future.

ABOUT HS PRESS

HS Press is an imprint of IRH Press Co., Ltd. IRH Press Co., Ltd., based in Tokyo, was founded in 1987 as a publishing division of Happy Science. IRH Press publishes religious and spiritual books, journals, and magazines, and also operates broadcast and film production enterprises. For more information, visit *okawabooks.com*.

Follow us on:

Facebook: Okawa Books
Youtube: Okawa Books
Pinterest: Okawa Books
Instagram: OkawaBooks
Twitter: Okawa Books
Goodreads: Ryuho Okawa

NEWSLETTER

To receive book-related news, promotions, and events, please subscribe to our newsletter below.

irhpress.com/pages/subscribe

AUDIO / VISUAL MEDIA

YOUTUBE

PODCAST

Visit the above to learn more about Ryuho Okawa's books. Topics range from self-help, current affairs, spirituality, religion, and the universe.

www.ingramcontent.com/pod-product-compliance
Lightning Source LLC
Chambersburg PA
CBHW031326230926
51848CB00001B/112
* 9 7 8 4 8 2 3 3 0 4 8 1 1 *